Workbook

for

Levin and Fox

Elementary Statistics in Social Research

Eighth Edition

Prepared by

Felicia P. Wiltz
Northeastern University

Allyn and Bacon
Boston London Toronto Sydney Tokyo Singapore

A Pearson Education Company
160 Gould Street
Needham Heights, MA 02494

Internet: www.abacon.com

ISBN 0-205-31102-4

Printed in the United States of America

10 9 8 7 6 5 4 3 2 1 03 02 01 99

Contents

Chapter 1: *Why the Social Researcher Uses Statistics* 1

Chapter 2: *Organizing the Data* 8

Chapter 3: *Measures of Central Tendency* 27

Chapter 4: *Measures of Variability* 38

Chapter 5: *Probability and the Normal Curve* 49

Chapter 6: *Samples and Populations* 59

Chapter 7: *Testing Differences Between Means* 72

Chapter 8: *Analysis of Variance* 86

Chapter 9: *Nonparametric Tests of Significance* 96

Chapter 10: *Correlation* 109

Chapter 11: *Regression Analysis* 121

Chapter 12: *Nonparametric Measures of Correlation* 131

Answers 142

1 Why the Social Researcher Uses Statistics

Multiple-Choice Questions

1. Imagine you are a social scientist and are interested in studying unemployment. Unemployment in this case would be a(n)

   a. variable.
   b. aggregate.
   c. hypothesis.
   d. constant.

2. In your study of unemployment you decide to interview women who have been jobless for more than three months. Each woman is a(n)

   a. variable.
   b. aggregate.
   c. constant.
   d. unit of observation.

3. A researcher conducts a study about violence in which experimental subjects are shown several films, each containing different amounts of violence. He thinks that when people see more violent films they become more agitated, thus increasing their propensity to commit violent acts. This idea is called a(n)

   a. constant.
   b. dependent variable.
   c. theory.
   d. hypothesis.

4. In question #3, the films would be a(n)

   a. independent variable.
   b. hypothesis.
   c. constant.
   d. dependent variable.

5. A researcher is interested in investigating the effects of job salary on job performance. Job performance level is the:

   a. independent variable.
   b. hypothesis.
   c. constant.
   d. dependent variable.

6. What level of measurement is job salary (from question #5)?

a. nominal scale.
b. ratio scale
c. interval scale.
d. ordinal scale.

7. Jonathan Smith is interested in investigating the dynamics between college recruiters and college students during the interviewing process. To gather data, he will work along side a team of college recruiters during their college recruitment process. This is an example of which research strategy?

a. survey
b. participant observation
c. experiment
d. content analysis

8. Dr. Beverly James is conducting research on employee hiring policies of two corporations. She has been utilizing human resource procedural manuals and records of hiring done for the last two years. Which kind of research method has she used?

a. survey
b. participant observation
c. experiment
d. content analysis

9. Dr. Bursik is interested in studying the Ku Klux Klan in order to learn about its organizational structure and activities. He forms a relationship with some of its members and is asked to join the local chapter. Over a six month period he participates as anyone else would, except that he takes extensive notes about his observations and about his conversations with Klan members. This is an example of which research strategy?

a. survey
b. participant observation
c. experiment
d. content analysis

10. A researcher has decided to use participant observation as a research method. He will not disclose his identity and therefore is using which type of participant observation?

a. content
b. covert
c. open
d. none of the above

11. A research team has been hired to evaluate the effect of a teen pregnancy prevention program on the views teens hold about pre-marital sex. The team measured teen views and then randomly selected two groups of teens. One group participated in the program and the other group did not participate. The research team then measured the teen views again. What kind of research method was used?

a. survey
b. content analysis.
c. experiment
d. all of the above.

12. Which research method does not directly observe behavior or question respondents?

a. survey
b. content analysis
c. experiment
d. all of the above

13. Dr. Caleb Young wants to find out if his patients approve of a new medical technique he is using. He has prepared a questionnaire to mail to his patients. The research method used here is:

a. survey
b. participant observation
c. experiment
d. content analysis

14. The first stage of social research is to

a. develop an appropriate set of measurement instruments.
b. reduce the problem to a testable hypothesis.
c. analyze data.
d. collect data.

15. What stage of social research follows data collection?

a. data gathering
b. instruments developed to gather data
c. formulate research problem
d. data analysis

16. The numbers on the backs of football players' jerseys are example of scores on a(n)

a. nominal scale.
b. ratio scale
c. interval scale.
d. ordinal scale.

17. Compared to the ordinal level of measurement, the interval scale

a. not only indicates the order of categories, but also the exact distance between them.
b. provides labeling of each score.
c. starts from a true zero point.
d. only categorizes.

18. Which of the following is true?

a. Common sense provides enough information about human behavior that formal hypothesis testing is rarely needed in the behavioral sciences.
b. Surveys are not often used to generalize beyond the sample available for study.
c. Experiments are not often used because they are inherently unethical.
d. The participant observer is able to determine the meaning of social situations as defined by group members themselves.

19. Creating a distribution of the incomes of all the people in a particular city is an example of using statistics as a tool of

a. decision making.
b. description.
c. surveys
d. hypothesis testing.

20. Generalizing the results of an experiment conducted on a small group of people to a larger group of people is an example of using statistics as a tool of

a. decision making.
b. description.
c. participant observation.
d. hypothesis testing.

21. Statistics can be used to

a. reduce data to more easily understood descriptive terms.
b. generalize results.
c. determine when an observed difference between two or more groups is the result of chance, or when it is the result of "real" differences between groups.
d. all of the above.

22. Which of the following "rules of thumb" are useful to remember when rounding?

a. Always round decimals down to the nearest whole number.
b. Always round decimals up to the nearest whole number.
c. Round a final answer to two more decimal digits than contained in the original scores.
d. You should round during all of the intermediate steps involved in arriving at an answer.

Exercises

23. A researcher was hired by a toy manufacturer to study the effects of noise levels on worker productivity. A very loud machine was allowed to run while workers affixed plastic eyes to the heads of dolls. At another time, no machines were allowed to run. In both cases, the number of eyes completed was recorded.

    a. What is the independent variable? _________

    b. What is the dependent variable? _________

24. One hundred people were asked their age and income. It was found that a person's age was related to their income level.

    a. What research strategy was used here? _________

    b. What is the independent variable? _________

    c. What is the dependent variable? _________

25. A researcher was interested in the behavior of adolescents in gangs. Specifically, he wondered how characteristics of each member's home life (parent's marital status, number of siblings, family income, etc.) influenced the leadership roles they assumed. The researcher recruited 30 gang members and asked them to record relevant aspects of their home life and the kinds of leadership roles that they have assumed.

    What research strategy was used here? _________

26. Dr. Heliotis is interested in studying how depression in mothers affects the emotional development of their young children. Samples of depressed and non-depressed mothers are identified by their scores on the Beck Depression Inventory. The children of these mothers are then observed by trained researchers who analyze how the children respond to three contrived situations, each lasting two minutes: (a) The mother stares at the child, remaining as still as possible and not interacting in any way. (b) The mother interacts with the child as she would normally. (c) The child is left alone. Dr. Heliotis expected the children of depressed mothers to be agitated in situation 'a', and less disturbed in situation 'b', than children of non-depressed mothers, since they should be used to less interaction from them.

    a. What research strategy was used here? _________

    b. What is the independent variable? _________

    c. What is the dependent variable? _________

27. A research team was hired to determine whether a rehabilitation program reduces the likelihood that prisoners will re-offend after they are released. Four hundred inmates were declared eligible for the program, and half were randomly assigned to it while the others are kept in prisons without participating in any rehabilitative programs. All of the inmates were studied for two years after their release to see whether they were arrested again. A comparison of arrest rates was made between the rehabilitation program and the conventional prison groups.

a. What research strategy was used here? _________

b. What is the independent variable? _________

c. What is the dependent variable? _________

28. A psychologist thinks that a major determinant of cognitive ability is an infant's diet, and he develops a nutritionally balanced diet specially designed for babies. 50 premature babies are randomly divided into 2 groups of 25. For one group, the researcher contacts the parents, explains the purpose and goals of the diet, and provides instructions and financial support for its implementation. The other group receives no special treatment of any kind. Two years later each child is given an intelligence test. It is found that the children who received the special diet had a higher average test score than did the other group of children.

a. What research strategy was used here? _________

b. What is the independent variable? _________

c. What is the dependent variable? _________

29. Dr. Black is a sociologist interested in studying gender differences in children's reactions to their parents' divorce. She hypothesizes that high school males will become more aggressive while females will not, and that this can be detected in their writing. To test her hypothesis, she goes to a high school and obtains papers written by 100 male and 100 female high school students, half of whose parents are divorced. She then examines each students' work for the presence of themes and language indicating aggression, and records the students' gender and parents' marital status.

a. What research strategy was used here? _________

b. What is the independent variable? _________

c. What is the dependent variable? _________

For questions 30 & 31, consider the sample survey question below

Q16. What is your marital status?

1 Married
2 Divorced or legally separated
3 Widowed

30. From the sample survey question above, what level of measurement is this variable?

____

31. Given the level of measurement of the variable in the sample survey question above, what criteria is not satisfied?

____

2 Organizing the Data

Summary of Equations

Proportion:

$$P = \frac{f}{N}$$

where f = number of cases in any given category
N = total number of cases in the distribution

Percentage:

$$\% = (100)\frac{f}{N}$$

where f = number of cases in any given category
N = total number of cases in the distribution

Ratio:

$$\text{Ratio} = \frac{f_1}{f_2}$$

where f_1 = frequency any given category
f_2 = frequency in any other category

Rate:

$$\text{Rate} = \frac{f_1}{f_2}$$

where f_1 = frequency of actual cases
f_2 = frequency of potential cases

Class Interval:

$$I = U - L$$

where i = size of a class interval
U = upper limit of a class interval
L = lower limit of a class interval

Midpoint:

$$m = \frac{\text{lowest score value} + \text{highest score value}}{2}$$

Cumulative Percentage:

$$c\% = (100)\frac{cf}{N}$$

where cf = cumulative frequency in any given category
N = total number of cases in the distribution

Percentile Rank for Grouped Frequency Distributions:

$$PR = c\%_b + \left(\frac{X - L}{i}\right)\%$$

where $c\%_b$ = cumulative percentage below the lower limit of the critical interval
X = raw score under consideration
L = lower limit of the critical interval
i = class interval size
% = percentage within critical interval

Cross-Tabulation:

A. ***Total***

$$\text{total\%} = (100)\frac{f}{N_{\text{total}}}$$

where f = the frequency in any category
N_{tot} = the total sample size

B. ***Row***

$$\text{row \%} = (100)\frac{f}{N_{\text{row}}}$$

where f = the frequency in any row
N_{row} = the number of cases in that row

C. ***Column***

$$\text{column \%} = (100)\frac{f}{N_{\text{column}}}$$

where f = the frequency in any column
N_{col} = the number of cases in that column

Multiple-Choice Questions

1. One way of standardizing frequency distributions (to compare groups of different total frequencies) is to use

   a. proportions.
   b. midpoints.
   c. intervals.
   d. cumulative distributions.

2. Proportions are multiplied by 100 in order to get

   a. intervals.
   b. percentages.
   c. midpoints.
   d. ratios.

3. Compared to proportions, ratios are used

   a. more commonly.
   b. less commonly.
   c. about as often.
   d. never in social research.

4. A researcher takes the number of cars sold by a certain dealer and divides this by the total number of cars shown by that same dealer. The resulting statistic is called a

   a. proportion.
   b. percentage.
   c. ratio.
   d. rate.

5. The term "fudge factor" is related to the concept of the

   a. proportion.
   b. percentage.
   c. midpoint.
   d. class interval.

6. Suppose that in a grouped frequency distribution, a class interval has a range of 100 to 110. The lower limit would be _____ and the upper limit would be _____.

   a. 100; 110
   b. 110; 100
   c. 99.5; 110.5
   d. 100.5; 109.5

7. The size of the class interval in question #6 is

   a. 10.
   b. 11.
   c. 5.
   d. 9.

8. The midpoint of the class interval in question #6 is

   a. 105.
   b. 104.5.
   c. 105.5
   d. 110.

9. Identifying the position of a score in an entire distribution of scores is best done using which of the following?

a. percentage.
b. class interval.
c. midpoint.
d. percentile rank.

10. Which of the following statements regarding class intervals is FALSE?

a. Each of the class intervals in a distribution must be of equal size.
b. Each class interval must have an upper and lower limit.
c. In constructing class intervals from a frequency distribution, there should not be too many or too few intervals.
d. In constructing class intervals, it is best to make the size of the interval a whole number rather than a decimal.

11. When constructing class intervals, which of the following are important guidelines?

a. The intervals must be mutually exclusive.
b. The intervals must be exhaustive.
c. Using too few intervals can result in a loss of precision, and obscure data patterns.
d. All of the above.

12. Cumulative frequencies are defined as

a. the total number of cases having a given score or a score that is higher.
b. the total number of cases having a given score or a score that is lower.
c. the percentage of cases having a given score or a score that is higher.
d. the percentage of cases having a given score or a score that is lower.

13. The percentile rank of a given score

a. indicates the percentage of scores in a distribution that fall at or below that score.
b. indicates the percentage of scores in a distribution that fall at or above that score.
c. indicates the proportion of scores in a distribution that fall at or below that score.
d. indicates the percentage of scores in a distribution that have the same score.

14. When dealing with class intervals with unbounded top or bottom class intervals,

a. you cannot estimate midpoints for these intervals.
b. you can precisely calculate the interval width using the proper formula.
c. you can use the width of the class interval next to the top or bottom interval to estimate their width.
d. none of the above.

15. A table presenting the distribution of one variable across the categories of one or more additional variables is called a

a. grouped frequency distribution.
b. cumulative frequency distribution.
c. cross-tabulation.
d. multivariate bar graph.

16. In a cross-tabulation, the clearest type of percentage to use (either row or column) is guided by the location of the

a. independent variable.
b. columns.
c. rows.
d. largest cell frequency.

17. A difference between pie charts and bar graphs is that the former

a. can accommodate any number of categories at any level of measurement.
b. are better at presenting more complex data.
c. are usually limited to showing the differences in frequencies or percentages of categories of a single nominal variable.
d. none of the above.

18. When constructing a bar graph, __________ data should be graphed using separated bars, while __________ or __________ data can be graphed with either joined or disjoined bars.

a. ordinal; nominal; ratio
b. nominal; ordinal; interval
c. ordinal; nominal; interval
d. interval; nominal; ordinal

19. A researcher is interested in young children's preferences for various toys. Children are observed playing with toys, and a rating is assigned to each of the toys available to the children. The rating scale ranges from 1 (children didn't like the toy) to 7 (children loved the toy). Which of the following graphic methods can this researcher NOT use?

a. bar graph
b. frequency polygon
c. pie chart
d. none of the above

20. A distribution is negatively when there are a few extreme __________ scores.

a. high
b. central
c. low
d. none of the above

21. Where is the tail of the distribution located in question #20?

a. left
b. right
c. center
d. none of the above

22. A perfectly symmetrical, unimodal distribution is known as

a. platypus.
b. mesokurtic.
c. katiecouric.
d. normal.

For the following data, answer questions 23 – 25.

70 – 77
62 – 69
54 – 61
46 – 53

23. What is the size of the class intervals?

a. 7
b. 6.5
c. 8
d. 7.5

24. What is the midpoint of the class interval 46 – 53?

a. 49
b. 49.5
c. 50
d. none of above

25. What is the lowest score value and the lower limit of the class interval 54 – 61?

a. 53, 54
b. 54.5, 53
c. 54, 53.5
d. none of the above

Exercises

26. A parents support group has 10 couples who participated in the program for 2, 3, 3, 3, 4, 6, 6, 9, 10, and 10 months. Develop a frequency distribution with cumulative frequencies and percentages.

27. With the frequency from question #26, answer the following:

a. What percentage of couples have been in the program less than 6 months?

____

b. How many couples have been in the program less than 10 months?

____

28. At a large university, the president made a controversial decision regarding a financial investment. A survey was administered to 1000 students - 500 undergraduates and 500 graduate students - asking for their opinion about the president's decision. The following table shows the frequencies of responses for the two groups:

	Class Status of Respondent	
Response	Undergraduate	Graduate
Agree with decision	165	80
Disagree with decision	240	322
Undecided	53	31
No Opinion	42	67
N =	500	500

Using the data above, answer the following questions:

a. What proportion of undergraduates disagreed with the president's decision?

____

b. What percentage of undergraduates responded "Undecided"?

____

c. What proportion of graduate students agreed with the president's decision?

____

d. What percentage of graduate students had no opinion regarding the president's decision? __________

29. Dr. Williams hands you the results of a quiz she has just given in class. There were 54 people taking the quiz. Below are the results.

Score	f	Score	f
25	2	12	3
24	1	11	1
23	2	10	2
22	2	9	3
25	3	8	1
20	4	7	0
19	5	6	0
18	5	5	0
17	6	4	1
16	4	3	0
15	4	2	0
14	3	1	0
13	2		

a. Construct a grouped frequency distribution of the above data, using 5 class intervals (from highest scores on the top to the lowest scores on the bottom of the distribution).

Class Interval	midpoint	*f*	%
Total		_____	_____

b. What is the class size of each interval? __________

c. Which class interval has the highest percentage of subjects in it? __________

d. What is the midpoint of the lowest class interval? __________

30. A professor collected the following data on how much money (in dollars) his students spent for supplies in his course. Construct a frequency table, with percentages, cumulative frequencies and cumulative percentages.

Student	Amount	Student	Amount
1	5	11	12
2	9	12	9
3	2	13	5
4	4	14	6
5	7	15	1
6	1	16	3
7	6	17	4
8	11	18	5
9	9	19	8
10	3	20	8

31. The following table was taken from a journal article. The data are based on an aggregate of 3 different communities in separate parts of the United States.

DIVORCE BY SOCIO-ECONOMIC STATUS (SES)

	SES		
	Low	Middle	High
Divorce	f	f	f
YES	40	59	23
NO	73	64	52
Total	113	123	75

a. What is the percentage of those in the low SES having a divorce? _____

b. What is the percentage of those in the high SES not having a divorce? _____

c. What is the proportion of those in the middle SES having a divorce? _____

d. What is the proportion of those in the low SES not having a divorce? _____

e. What is the percentage of those in the high SES having a divorce? _____

f. What is the proportion of those in the middle SES not having a divorce? _____

32. The following is a group frequency distribution of student scores for a math quiz worth 50 points.

Class Interval	frequency	Class Interval	Frequenc y
0 – 5	2	26 – 30	6
6 – 10	1	31 – 35	8
11 – 15	4	36 – 40	3
16 – 20	6	41 – 45	3
21 – 25	5	46 – 50	1

a. What is the size of the class intervals? __________

b. What is the midpoint of the class interval with the highest test score? __________

c. What are the upper and lower limits of the class interval with the highest frequency?

__________ and __________

d. What percentage of students scored above 40 on the quiz? __________

e. What percentage of students scored between 26 and 35? __________

33. A group of males and females were asked to provide their opinion of recent welfare reform legislation, on a scale from 1 (strongly disapprove) to 5 (strongly approve).

	Sex of Respondent		
Rating	Male	Female	Total
5	26	33	59
4	38	41	79
3	53	35	88
2	48	24	72
1	29	12	41
Total	194	145	339

a. What percentage of the sample are male and give a rating of 2? __________

b. What percentage are female and give a rating of 1? __________

c. What percentage gave a rating of 5? __________

d. What percentage of the respondents who gave a rating of 3 are female? __________

e. What percent of those respondents who gave a rating of 4 are male? __________

f. What percent of males gave a rating of 2? __________

g. What percent of females gave a rating of 3? __________

34. A group of consumers were asked to provide a satisfaction rating for a new brand of cereal, on a scale from 1 (strongly disapprove) to 5 (strongly approve). The results are presented by racial category, as follows:

Rating	Whites	Blacks	Hispanics	Asian
5	24	15	13	5
4	36	22	9	4
3	41	20	12	6
2	19	9	6	1
1	10	8	3	2

a. What percentage of **all** respondents gave a rating of 3? __________

b. What percentage of blacks gave a rating of 4? __________

c. What percentage of **all** respondents are non-white and gave a rating of 5?

____

d. What percentage of **all** respondents are white? __________

e. What percentage of **all** respondents are hispanic and gave a rating lower than 4?

____

35. Use the bar graph below to answer the following questions:

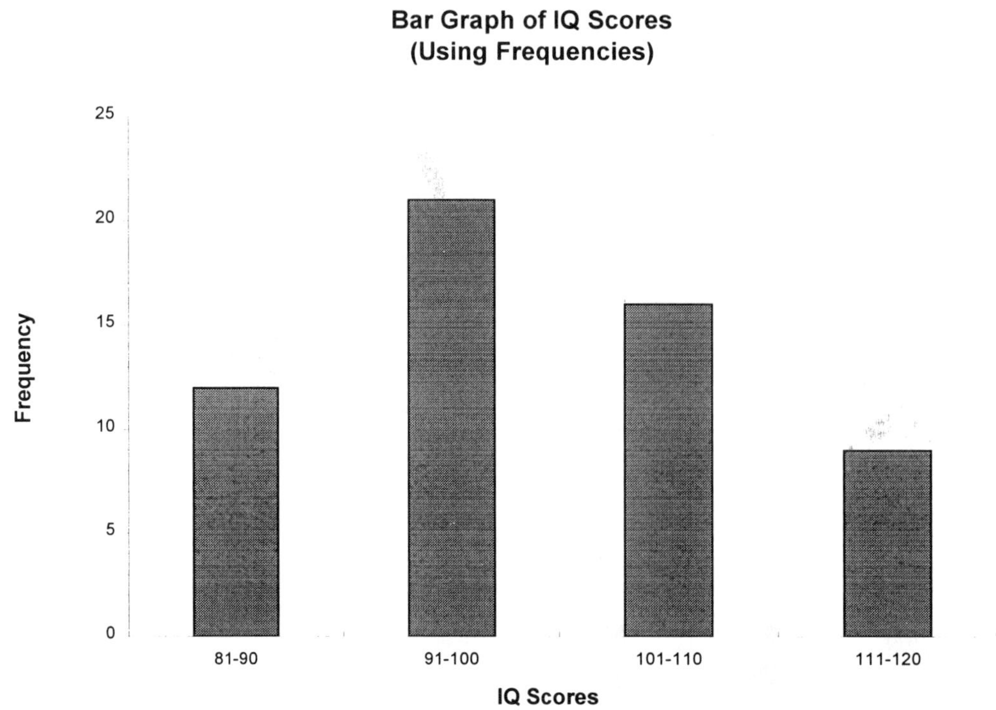

a. Which class interval contains the greatest frequency? __________

b. What is the frequency of individuals with IQs over 110? __________

c. What is the frequency of individuals with IQs below 91? __________

d. What is the midpoint of the second highest class interval? __________

36. Use the bar graph below to answer the following questions:

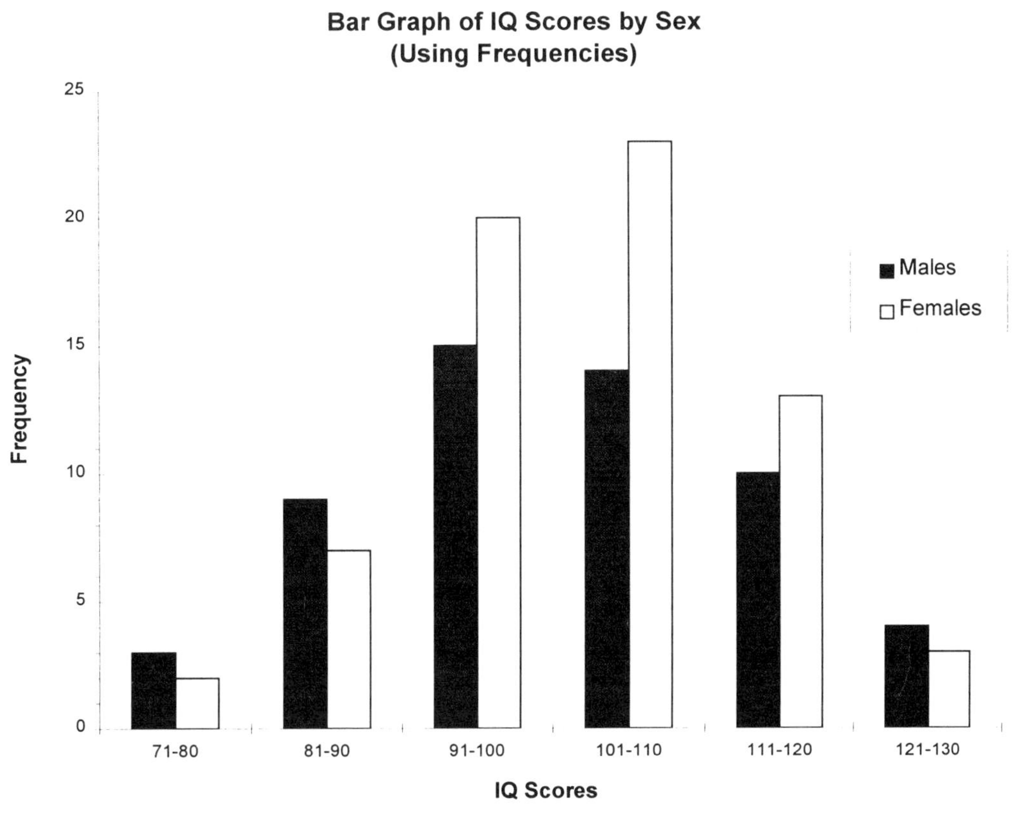

a. Which group has FEWER instances of IQs above 120? __________

b. How many females have IQs between 100 and 111? __________

c. What is the frequency of males with IQs below 91?

d. Use the graph to construct a grouped frequency distribution of IQ scores for females.

Class Interval	f	%	cf	c%
Total	______	______		

e. From this grouped frequency distribution, determine what percentage of females have IQs above 90. __________

37. The following is an ethnic breakdown of a local insurance company's employees:

Ethnic Background	f	%
Black	50	25.0
Hispanic	25	12.5
White	75	37.5
Other	50	25.0
Total	200	100.0

a. Construct a pie chart using the above data (using percentages).

b. Construct a vertical bar chart representing the above data (using frequencies).

38. A small college reports the following scores on an entrance examination for its incoming freshman class:

Class Interval	f	%	cf	c%
1400-1499	5	3.40	147	100.00
1300-1399	4	2.72	142	96.60
1200-1299	9	6.12	138	93.88
1100-1199	12	8.16	129	87.76
1000-1099	13	8.84	117	79.59
900-999	13	8.84	104	70.75
800-899	15	10.20	91	61.90
700-799	21	14.29	76	51.70
600-699	17	11.56	55	37.41
500-599	12	8.16	38	25.85
400-499	9	6.12	26	17.69
300-399	6	4.08	17	11.56
200-299	4	2.72	11	7.48
100-199	5	3.40	7	4.76
0-99	2	1.36	2	1.36
Total	147	100.00		

a. Construct a vertical bar chart of the above distribution, using frequencies.

b. Construct a frequency polygon of the above distribution.

c. Construct a cumulative frequency polygon of the above distribution.

3 Measures of Central Tendency

Summary of Equations

Median:

A *Position of median in ordered array of raw data*

$$\text{Position of median} = \frac{N + 1}{2}$$

B *Estimating median from grouped frequency distribution*

$$\text{Median} = L + \left(\frac{N/2 - cf_b}{f} \right) i$$

where

N	=	number of cases in distribution
cf_b	=	cumulative frequency below the lower limit of the critical interval
L	=	lower limit of the critical interval
f	=	frequency of the critical interval
i	=	class interval size

Mean:

A *Raw data*

$$\overline{X} = \frac{\Sigma X}{N}$$

where			
	$\overline{X}$	=	the mean
	Σ	=	the sum
	X	=	a raw score in the set of scores
	N	=	the total number of scores in the set

B *Frequency distribution*

$$\overline{X} = \frac{\Sigma fX}{N}$$

where			
	$\overline{X}$	=	the mean
	X	=	a raw score in the distribution
	f	=	the frequency of occurrences of X
	N	=	the total number of scores in the set

C *Grouped frequency distribution*

$$\overline{X} = \frac{\Sigma fm}{N}$$

where			
	$\overline{X}$	=	the mean
	m	=	the midpoint of a class interval
	f	=	the frequency of a class interval
	N	=	the total number of scores in the set

Deviation:

deviation $= X - \overline{X}$

where			
	X	=	any raw score in the distribution
	$\overline{X}$	=	the mean of the distribution

Multiple-Choice Questions

1. A measure of central tendency is

   a. the deviation.
   b. a variable.
   c. the distribution.
   d. the mode.

2. A number that represents what is "average" or "typical" in a particular data set is the

   a. central tendency.
   b. mean.
   c. mode.
   d. median.

3. The most frequent, typical, or common value in a distribution is called the

   a. mean.
   b. deviation.
   c. mode.
   d. median.

4. A researcher who has data with modes of 14 and 25 has what kind of distribution?

   a. bimodal
   b. skewed
   c. unimodal
   d. symmetrical

5. The middle point in a distribution is the

   a. mean.
   b. deviation.
   c. mode.
   d. median.

6. If you were to randomly select 20 social science research articles you would probably find the __________ to be the most commonly used measure of central tendency.

   a. mode
   b. median
   c. deviation
   d. mean

7. The type of central tendency used for nominal variables is

   a. mean.
   b. median.
   c. mode.
   d. midpoint.

8. The deviation

   a. may be regarded as the "center of gravity" of a distribution.
   b. is the middle-most point in a distribution
   c. indicates the direction and distance or any raw score from the mean.
   d. indicates the direction and distance or any raw score from the median.

9. As part of a class project, you hand out a questionnaire to 100 students. One of the questions attempts to classify individuals on the basis of race (e.g., African-American, Caucasian, Asian, Latino, etc.). Which of the following measures of central tendency may be used to describe the data for this question?

a. mean
b. mode
c. median
d. all of the above

10. In a study of interpersonal attraction a social psychologist presents subjects with 10 photographs of people, and asks them to order the photos from most to least pleasing. Which of the following measures of central tendency may be used to describe the resulting data?

a. mean
b. mode
c. median
d. all of the above

11. Melissa Daniels has compiled a list of starting salaries for a particular position. Which measure of central tendency can be applied?

a. mean
b. mode
c. median
d. all of the above

12. Which measure of central tendency is very influenced by the extreme scores that may appear in a skewed distribution?

a. mean
b. mode
c. median
d. all of the above

13. Which of the following is an accurate statement about the distribution below?

a. The ordering from low to high scores of measures of this distribution's central tendency is mean, median, then mode.
b. The ordering from low to high scores of measures of this distribution's central tendency is median, mean, then mode.
c. The ordering from low to high scores of measures of this distribution's central tendency is mode, mean, then median.
d. The mean, median, and mode are all positioned identically.

14. From left to right in the distribution below, the order of the measures of central tendency is

a. mode, median, mean
b. median, mode, mean
c. mode, mean, median
d. mean, median, mode

15. Which of the following measures would best describe the central tendency the distribution in question #14?

a. mean
b. median
c. mode
d. all of the above

16. A group containing 28 rankings of 10, 26 rankings of 9, and 20 rankings of 8, has a mode of _____.

a. 28
b. 9
c. 26
d. 10

17. With the scores 5, 8, 12, 2, 5, 7, and 9, the mode is

a. unimodal
b. 12
c. 7
d. bimodal

18. Imagine you are doing a study of social status and one of your measures of a family's standing is the number of bedrooms in their home. The first 10 families studied had 3, 1, 2, 2, 4, 3, 1, 4, 2 and 9 bedrooms. You wish to report the central tendency of your distribution of the number of bedrooms per family home. Which of the following measures would be less sensitive to the presence of the 9 bedroom home.

a. mean
b. median
c. average
d. none of the above

19. You go to an all-you-can-eat Italian restaurant, and you and your friends count the number of plates of spaghetti you each ate: 1, 2, 2, 3, 8. At this restaurant, any table that eats "*on average*" three or more plates of spaghetti wins a $10 coupon. You and your friends want to win this coupon. Given your individual spaghetti-eating results, which of the three types of "average" or central tendency would you calculate and report to the manager? Given that the manager wants to avoid giving the coupon, which of the three measures would she want to use as your table's average?

a. You and your friends would use the mean, and so would the manager.
b. You and your friends would use the mean, and the manager would use the median.
c. You and your friends would use the median, and the manager would use the mean.
d. You and your friends would use the median, and so would the manager.

Exercises

20. Indicate the relative positions from left to right of the MEAN, MEDIAN, and MODE in the following distribution. _____ _____ _____

21. Indicate the relative positions from left to right of the MEAN, MEDIAN, and MODE in the following distributions. _____ _____ _____

22. Indicate the relative positions from left to right of the MEAN, MEDIAN, and MODE in the following distributions. _____ _____ _____

23. Twenty-three economics students were hired during the senior job fair. Below is the frequency distribution of their grade-point averages.

Grade Point Average	f
3.9	1
3.7	3
3.5	1
3.4	6
3.1	5
3.0	4
2.9	3
2.6	4
2.4	2

a. What is the mode? _____

b. What is the median? _____

c. What is the mean? _____

24. An insurance company's college work study program has recorded the number of hours eacy student worked last week. The hours were: 24; 16; 20; 24; 19; 10; 18; 15.

a. What is the mode? _____

b. What is the median? _____

c. What is the mean? _____

25. Jean Eaton, a human resources manager, is interested in the number of employees terminated for reasons of absenteeism during the past year. She spoke to departmental supervisors to find out how many infractions each employee had prior to termination. The results were:

Former Employee	Infractions
Richard	8
Lyndsay	6
Colleen	9
Bob	4
Paul	7
Kris	7
David	8

a. What is the mean? _____

b. What is Colleen's deviation from the mean? _____

c. What is Lyndsay's deviation from the mean? _____

26. The Dress Shoppe surveyed their regular customers to find out their ages. These are the results:

35	37	38	29	30	33	38	42
40	30	29	32	33	38	31	29
32	37	36	41	33	36	39	35

a. What is the mean age? _____

b. What is the modal age? _____

c. What is the median age? _____

d. What is the deviation from the mean of an age of 42? _____

27. Given the following scores, answer the questions below.

98, 94, 90, 89, 85, 76, 74, 74, 72, 67, 64

a. Given a score of 91, what is the deviation from the mean? _____

b. Given a score of 75, what is the deviation from the mean? _____

c. Given a score of 64, what is the deviation from the mean? _____

28. A 100-point multiple choice exam is given in a class. Below is a distribution of scores:

Class Interval	Midpoint	Frequency	cf	%
91-100	95.5	2	80	4.00
81-90	85.5	7	48	14.00
71-80	75.5	6	41	12.00
61-70	65.5	8	35	16.00
51-60	55.5	9	27	18.00
41-50	45.5	8	18	16.00
31-40	35.5	4	10	8.00
21-30	25.5	3	6	6.00
11-20	15.5	2	3	4.00
1-10	5.5	1	1	2.00
		N = 50		

a. What is the mode? _____

b. What is the median? _____

c. What is the mean? _____

29. Dr. Lam is a criminologist interested in the level of compliance of people ordered to perform community service as a condition of their probation. She selects a sample of 9 people, each of whom were sentenced to perform 50 hours of service. She contacts each offenders' work site supervisor and obtains records of the total number of hours actually completed. The results were:

Offender's Name	Hours of Community Service
Julie	23
Tom	41
John	31
Tim	14
Liz	45
Seamus	27
Carolyn	34
Leueen	39
Gary	25

a. What is the mean number of hours of community service performed? _____

b. What is the deviation from the mean of Julie's hours performed? _____

c. What is the deviation from the mean of John's hours performed? _____

d. What is the deviation from the mean of Tim's hours performed? _____

30. The approval ratings for student interns are listed below. (1 represents very low and 10 represents very high)

Student	Rating
Gina	3
Chris	9
Mike	10
Marie	4
Sean	9
Gabe	5
Andrea	10
Ronnie	7
Julia	8

a. What is the modal rating? _____

b. What is the median rating? _____

c. What is the mean rating? _____

d. What is the deviation from the mean of Gina's score? _____

e. What is the deviation from the mean of Gabe's score? _____

31. A welfare program in Detroit has a warehouse containing packages of government surplus cheese. The cheese comes in 1/4, 1/2, 1, and 2 pound packages. The manager of this program does inventory and creates a frequency distribution of the packages in stock:

Size of Package (in pounds)	Number of Packages
1/4	100
1/2	500
1	50
2	20

a. What is the mode of this frequency distribution? _____

b. What is the deviation from the mean of a 1-pound package? _____

32. Forty students have been accepted to the a graduate school of economics. Below is a frequency distribution of their grade-point averages:

GPA	Number of Students
2.0	2
2.3	3
2.6	7
2.9	11
3.2	7
3.5	7
3.8	3

a. What is the modal GPA? _____

b. What is the median GPA? _____

c. What is the mean GPA? _____

33. Below are the ages of a group of people applying for jobs as research assistants at the Center for the Study of Social Issues:

20, 20, 20, 20, 21, 21, 21, 21, 21, 21,
21, 22, 22, 22, 22, 22, 22, 22, 22, 22,
22, 22, 22, 22, 22, 22, 23, 23, 23, 23,
23, 23, 23, 23, 23, 23, 23, 23, 23, 24,
24, 24, 24, 24, 24, 24, 24, 25, 25, 25

a. What is the mean age? _____

b. What is the median age? _____

c. How much does the age of a 24 year old deviate from the mean? _____

4 Measures of Variability

Summary of Equations

Range:

R = H - L

where			
where	H	=	highest score in a distribution
	L	=	lowest score in a distribution

Mean Deviation:

$$MD = \frac{\Sigma |X - \bar{X}|}{N}$$

where			
where	MD	=	mean deviation
	$\Sigma\lvert X - \bar{X}\rvert$	=	sum of the absolute deviations (disregarding plus and minus signs)
	N	=	total number of scores

Variance:

A. *Definitional formula*

$$s^2 = \frac{\Sigma(X - \bar{X})^2}{N}$$

where			
where	s^2	=	variance
	$\Sigma(X - \bar{X})^2$	=	sum of the squared deviations from the mean
	N	=	total number of scores

B. *Raw-score formula*

$$s^2 = \frac{\Sigma X^2}{N} - \overline{X}^2$$

where ΣX^2 = sum of the squared raw scores (each raw score is first squared, and then these squared raw scores are added)

N = total number of scores

$\overline{X}^2$ = mean squared

C. *Simple frequency distribution formula*

$$s^2 = \frac{\Sigma fX^2}{N} - \overline{X}^2$$

where X = score values

f = score frequency

N = number of cases

$\overline{X}$ = mean of the simple frequency distribution

D. *Grouped frequency distribution formula*

$$s^2 = \frac{\Sigma fm^2}{N} - \overline{X}^2$$

where m = class interval midpoints

f = class interval frequency

N = number of cases

$\overline{X}$ = mean of the grouped frequency distribution

Standard Deviation:

A. *Definitional formula*

$$s = \sqrt{\frac{\Sigma(X - \overline{X})^2}{N}}$$

where s = standard deviation

$\Sigma(X - \overline{X})^2$ = sum of the squared deviations from the mean

N = total number of scores

B. *Raw-score formula*

$$s = \sqrt{\frac{\Sigma X^2}{N} - \overline{X}^2}$$

where ΣX^2 = sum of the squared raw scores (each raw score is first squared, and then these squared raw scores are added)

N = total number of scores

$\overline{X}^2$ = mean squared

C. *Simple frequency distribution formula*

$$s = \sqrt{\frac{\Sigma f X^2}{N} - \overline{X}^2}$$

where X = score values

f = score frequency

N = number of cases

$\overline{X}$ = mean of the simple frequency distribution

D. *Grouped frequency distribution formula*

$$s = \sqrt{\frac{\Sigma f m^2}{N} - \overline{X}^2}$$

where m = class interval midpoints

f = class interval frequency

N = number of cases

$\overline{X}$ = mean of the grouped frequency distribution

Multiple-Choice Questions

1. Which of the following is NOT a measure of variability? (2 points)

   a. Variance
   b. Standard Deviation
   c. Range
   d. Mean

2. The spread or dispersion of a distribution is commonly referred to as

   a. the standard deviation.
   b. the range.
   c. the variance.
   d. variability.

3. Which of the following measures of variability is no longer widely used by social science researchers?

   a. standard deviation
   b. range
   c. mean deviation
   d. variance

4. On a scale designed to measure the positive social adjustment of children, a social worker reports that the highest score obtained in the latest tests was 98 (out of 100) while the lowest was 32. What measure of variability has he reported?

   a. standard deviation
   b. range
   c. variance
   d. mean deviation

5. Given the information in question #4, what can you confidently conclude about the distribution of social adjustment scores?

   a. Most of the scores probably fall near or around 70.
   b. Most of the scores are probably closer to 98 than to 32.
   c. Few of the scores are probably less than 50, the midpoint of the scale.
   d. It is a skewed distribution.
   e. None of the above.

6. The difference between the standard deviation and the variance is that

   a. the standard deviation is larger.
   b. the standard deviation represents deviation in the original unit of measurement, while the variance is inflated due to squaring.
   c. the variance uses absolute values in calculating the deviations.
   d. the standard deviation is a measure of variability, while the variance is a measure of central tendency.

7. Distribution A has a standard deviation of 1.6 while distribution B has a standard deviation of 2.8. This indicates that

   a. distribution B has less variability than distribution A.
   b. distribution A has greater variability than distribution B.
   c. distribution B has greater variability than distribution A.
   d. it is impossible to tell which distribution has more variability without knowing the mean.

8. You plot the times that it takes a sample of rats to run a maze, and you notice that the distribution is relatively normal and symmetrical. What can you conclude about the scores?

   a. About one-fourth of the scores will fall within one standard deviation above and below the mean.
   b. The distribution of score is leptokurtic.
   c. About two-thirds of the scores will lie beyond one standard deviation above and below the mean.
   d. About one-third of the scores will lie beyond one standard deviation above and below the mean.

9. A distribution containing 500 scores has a high score of 55 and a low of 31. What would be the best guess about the standard deviation of this distribution.

   a. 24
   b. 12
   c. 4
   d. 1

10. A distribution has 28 scores, a mean of 16, and a standard deviation of 3. What is the variance of this distribution?

    a. 4
    b. 9
    c. 8
    d. can't tell (no raw data)

11. An instructor hands back exams and tells you that the mean score is 74, with a standard deviation of 10. A few days later, the instructor has noted that there were some mistakes in the exam, and she decides to give everyone an extra 3 points. The new standard deviation is

    a. 13
    b. 10
    c. 10
    d. can't tell (no raw data)

12. Which of the following sets of scores has the greatest variability?

    a. 10, 11, 12, 13, 14, 15
    b. 5, 12, 13, 19, 20, 21
    c. 5, 10, 15, 20, 25, 30
    d. 5, 15, 25, 35, 45, 55

13. Two new training programs were being implemented, and the people receiving the training were asked to rate its quality (on a 7-point scale, with 1 being "extremely low" and 7 being "extremely high"). Program #1 received rating scores of 1, 3, 1, 5, 2, 6, 7, 7, 4, 2. Program #2 received scores of 4, 3, 3, 4, 5, 5, 6, 4, 4, 5. From this, you could conclude that

a. trainees liked Program #1 much better than Program #2.
b. there was greater variability in the reactions to Program #2 than in reactions to Program #1.
c. there was greater variability in the reactions to Program #1 than in reactions to Program #2.
d. trainees reacted very similarly to both programs.

14. In calculating the mean deviation, each score's difference from the mean is summed using the absolute value (so that if the mean were 52 and a score was 48, the deviation would be 4 rather than -4). If the plus and minus signs were used, the deviations would sum to

a. zero.
b. a negative number.
c. a positive number.
d. the mean.

15. In calculating the mean deviation, use of the absolute value in determining the sum of deviations means that

a. as the sum of the absolute deviations becomes larger, the variability of a distribution tends to increase.
b. as the sum of the absolute deviations becomes larger, the variability of a distribution tends to decrease.
c. the sum of absolute deviations will always be centered around 0.
d. greater emphasis is given to extreme scores.

16. In calculating the mean deviation, absolute values are used in summing the deviations in a distribution. In calculating the variance, the problems of positive and negative deviations is avoided by

a. taking the square root of deviations.
b. using absolute values.
c. squaring the deviations.
d. ignoring extreme scores.

17. Difficulty interpreting the unit of measurement is associated with the

a. mean deviation.
b. variance.
c. standard deviation.
d. range.

Exercises

18. Compute the following given these scores:

25 16 19 14 23 13 21 10 11

a. Range: _____

b. Variance: _____

c. Standard deviation: _____

19. Compute the following given these scores:

3 9 7 3 6 4 1 8 9 3

a. What is the range for this distribution? _____

b. What is the variance for this distribution? _____

c. What is the standard deviation for this distribution? _____

20. A sleep analysis researcher recorded the number of hours of sleep 14 participants had after drinking a special herbal tea.

Males	**Females**
6	5
8	7
5	8
3	4
7	6
6	4
9	7

a. What is the mean number of hours of sleep for each group?

_____ _____

b. What is the range for each group? _____ _____

c. What is the mean deviation for each group? _____ _____

d. What is the standard deviation for each group? _____ _____

e. Which group has the greatest variability? __________

21. The number of hours a group of individuals spent in traffic last week are 2, 2, 3, 4, 4, 5, 8, and 9. Compute the following:

a. What is the range for this distribution? _____

b. What is the variance for this distribution? _____

c. What is the standard deviation for this distribution? _____

22. A researcher studying pro-social behavior devised the following experiment: A female confederate (part of the research team, but posing as just another pedestrian) walking along a busy sidewalk pretends to trip and fall, and lays on the ground in mock pain until someone offers to help her. Another researcher records how many seconds it takes for someone to stop to assist her. This process is repeated many times, and the data obtained are presented in the following grouped frequency distribution:

Class Interval	midpoint	f	%
21-25	23	1	1.85
16-20	18	6	11.11
11-15	13	13	24.07
6-10	8	10	18.52
1-5	3	24	44.44
Total		54	100.00

What is the standard deviation for this distribution? _____

23. The number of cups of coffee a group of workers had in one day are 0, 0, 1, 2, 2, 3, and 4. Compute the following:

a. What is the range for this distribution? _____

b. What is the standard deviation? _____

24. For this set of scores, 7, 6, 4, 2, 1, 1, 0, compute the following.

a. What is the range? _____

b. What is the mean deviation? _____

25. You need to take a history course in order to graduate. Below are the grades given last semester by the professor who teaches this course (the grades are expressed as whole numbers, e.g. 4 = A, 3 = B, etc.):

3	4	3
2	2	4
1	1	2
4	4	3
3	3	4
3	2	4
3	3	3
4	4	4
3	2	2
2	2	1

a. What is the mean grade given by this professor? _____

b. What is the standard deviation of the grade distribution? _____

26. The starting salaries for Research Assistants in two different companies are listed below (in thousands of dollars). The companies are very similar in terms of number of employees, net worth, health benefits, etc.

Social Policy Research, Inc.	Center for Social Research
22	15
25	18
27	23
23	32
24	28
26	25
21	19
24	30

a. What is the range for Social Policy Research, Inc.? _____

b. What is the range for the Center for Social Research? _____

c. What is the mean starting salary for each company? _____

d. If you wanted to be sure of making around $24,000 per year, which company should you probably choose? __________ Why? ______________________________

e. If you thought you had superior skills and felt you deserved to be rewarded with as much money as possible, which company should you probably choose? __________ Why? ______________________________

The following is a record of overtime hours both Jan and Peter worked each week for the past five weeks. Using this data, answer questions 27 – 30.

Jan	Peter
5	3
2	4
7	8
0	4
5	3

27. Calculate the range for Jan and for Peter. _____ _____

28. Calculate the mean deviation for Jan and for Peter. _____ _____

29. Calculate the standard deviation for Jan and for Peter. _____ _____

30. Who has greater variability of overtime hours? __________

5 Probability and the Normal Curve

Summary of Equations

Probability:

$$\text{probability of an outcome or event} = \frac{\text{number of times the outcome or event can occur}}{\text{total number of times any outcome or event can occur}}$$

***z* score**:

$$z = \frac{X - \mu}{\sigma}$$

where μ = mean of a distribution
σ = standard deviation of a distribution
z = a standard score

Multiple-Choice Questions

1. Testing hypotheses through the analysis of data is also known as

a. probability.
b. descriptive statistics.
c. decision making.
d. standardization.

2. Which of the following is NOT a probability?

a. 1
b. .08
c. 12
d. 100%

3. One minus the probability of an event occurring is also known as the

a. converse rule.
b. multiplication rule.
c. mutual exclusiveness rule.
d. addition rule

4. A bag is filled with red, white, and black marbles. You can only draw one marble out of the bag at a time. If you tried to determine the probability of drawing either a red or a white marble on the first draw, you would be using the

a. converse rule.
b. multiplication rule.
c. mutual exclusiveness rule.
d. addition rule.

5. The __________ rule assumes that each event or outcome is mutually exclusive.

a. converse
b. multiplication
c. addition
d. subtractive

6. Suppose you had two bags of red, white, and black marbles, and you drew a marble out of each bag. Now if you tried to determine the probability of drawing a red marble out of one bag and a black marble from the other, you would be using the

a. converse rule
b. multiplication rule
c. mutual exclusiveness rule
d. addition rule

7. Each of two bags is filled with red, white, blue, and black marbles. The probability of picking a blue marble from one bag is .30, and the probability of picking a red marble from the other bag is .45. What is the probability of drawing a blue marble from one bag and then a red marble from the other?

a. .75
b. .135
c. .15
d. .768

8. If there are 10 people in a room and 3 of them are men, what is the probability that the next person leaving the room will be a woman?

a. .30
b. 70%
c. .70
d. 1/3

9. The sum of all possible outcomes of an event (assuming mutually exclusive outcomes) is

a. .50.
b. the number of possible outcomes.
c. 1.
d. .25.

10. Suppose that you were to toss a single coin 4 times. What is the probability that all 4 tosses would result in "heads"?

a. .125
b. .20
c. .0625
d. .16

11. In rolling a single die, the probability of getting an odd number is

a. 1/6.
b. 3/6.
c. 2/6.
d. 1/2.

12. The mean of a probability distribution is

a.
b. s.
c. σ
d. μ

13. A perfectly symmetrical, unimodal distribution is known as the

a. ojive curve.
b. bell-jar curve.
c. normal curve.
d. regression curve.

14. All distributions in the social sciences conform to the normal curve.

a. True.
b. False.

15. An instructor hands back tests to a very large class (200 students). She says that the distribution of the test scores was exactly normal, with a mean of 50 (out of 100) and a standard deviation of 10 points. Approximately how many students scored between 50 and 60 (inclusive)? (Hint: You don't need to calculate a z score to answer this question.)

a. 100
b. 68
c. 50
d. 136

16. Getting heads or tails on a coin toss are two events that are mutually exclusive and exhaustive. What is the probability of getting either heads or tails on the tenth coin toss?

a. 1
b. .10
c. 100
d. .001

17. A college course contains 6 freshmen, 10 sophomores, 4 juniors, and 5 seniors. If you were to randomly select 1 person from this course, what is the probability that you would select either a freshman or a sophomore?

a. 10/25
b. 16/25
c. 15/25
d. 9/25

18. A research firm has been contracted to study people's preferences in music. Five different music samples (heavy metal, reggae, country, classical, and rap) were given to subjects, who rated how much they liked each. If we assume that people have NO preferences to different types of music (that is, everyone rates each type equally), what proportion of the time would classical music be rated highest?

a. 1/2
b. 1/5
c. 1/25
d. 1/625

19. In a standard deck of cards, the probability of drawing a spade is

a. 1/13.
b. 13/52.
c. 4/52.
d. 1/26.

20. In a standard deck of cards, the probability of drawing a face card (jack, queen, or king) is

a. 1/13.
b. 1/52.
c. 12/52.
d. 12/26.

21. The mean salary for the clerical staff at Company X is $22,000 with a standard deviation of $1250. How many standard deviations is the salary of $24,125 above the mean?

a. 2.8
b. 1.7
c. 0.7
d. 2.3

22. Given the data in question #21, how many standard deviations is the salary of $21,050 below the mean?

a. 0.67
b. 1.05
c. 0.76
d. 1.76

23. Given the data in question #21, how many standard deviation is the salary of $18,750 below the mean?

a. 2.5
b. 1.6
c. 2.7
d. 2.6

24. The numbers listed in column (a) of Table A in the text express the

a. proportion of the normal curve less than a given *z* score.
b. proportion of the normal curve less than a given raw score.
c. the percentage of a normal curve between the mean and a given *z* score.
d. the percentage of a normal curve between the mean and a given raw score.

25. The numbers listed in column (b) of Table A in the text express the

a. proportion of the normal curve beyond a given *z* score.
b. proportion of the normal curve beyond a given raw score.
c. the percentage of a normal curve beyond a given *z* score.
d. the percentage of a normal curve beyond a given raw score.

26. Fifty percent of the area of the normal curve lies below the mean, so if you were to randomly select one score from the distribution

a. there is a 50% chance that you would draw a score below the mean.
b. there is a probability of .5 that you would draw a score above the mean.
c. both a and b.
d. none of the above.

27. Suppose you have a normal distribution of IQ scores, with a mean of 100 and a standard deviation of 15. If you were to randomly select 1 score from this distribution, the probability that you would draw a score less than 85 is __________ the probability of drawing a score greater than 115.

a. less than
b. greater than
c. equal to
d. none of the above

28. Again, suppose you have a normal distribution of IQ scores, with a mean of 100 and a standard deviation of 15. If you were to randomly select 1 score from this distribution, the probability that you would draw a score less than 85 is __________ the probability of drawing a score greater than 100.

a. less than
b. greater than
c. equal to
d. none of the above

29. Again, suppose you have a normal distribution of IQ scores, with a mean of 100 and a standard deviation of 15. If you were to randomly select 1 score from this distribution, the probability that you would draw a score less than 100 is __________ the probability of drawing a score greater than 100.

a. less than
b. greater than
c. equal to
d. none of the above

30. Again, suppose you have a normal distribution of IQ scores, with a mean of 100 and a standard deviation of 15. If you were to randomly select 1 score from this distribution, the probability that you would draw a score less than 85 is

a. .1587.
b. .3413.
c. 68.13 %.
d. .6813.

31. Again, suppose you have a normal distribution of IQ scores, with a mean of 100 and a standard deviation of 15. If you were to randomly select 1 score from this distribution, the probability that you would draw a score greater than 130 is

a. 4.56%.
b. .4772.
c. .0228.
d. .0456.

32. Again, suppose you have a normal distribution of IQ scores, with a mean of 100 and a standard deviation of 15. If you were to randomly select 1 score from this distribution, the probability that you would draw a score between 85 and 115 is

a. .6826.
b. .3413.
c. .1587.
d. .3174.

Exercises

33. A bag is filled with 3 white, 5 blue, and 6 green marbles.

a. What is the probability of drawing a green marble from the bag with one draw?

_____

b. Express the probability of NOT drawing a white marble with one draw: _____

c. What is the probability of drawing either a white or a blue marble with one draw?

_____

d. What is the probability of drawing either a white, a blue, or a green marble? _____

34. What is the probability of rolling a single die and having it come up "3" four times in a row? _____

35. Given a standard deck of cards:

a. What is the probability of drawing either the king of spades or the ace of diamonds?

_____

b. What is the probability of NOT drawing a red card? _____

c. What is the probability of drawing a red 6? _____

36. In rolling two dice, let X represent the total shown by the dice. For example, the probability of the dice totaling 2 is shown below:

$$
\begin{aligned}
P(2) &= P(1 \text{ on one die})P(1 \text{ on another die}) \\
&= (1/6)(1/6) \\
&= 1/36
\end{aligned}
$$

The probability of the dice totaling 3 is shown below (the first die can show a 1 and the second a 2, or the first can show a 2 and the second a 1):

$$
\begin{aligned}
P(3) &= P(1,2 \text{ or } 2,1) \\
&= P(1,2) + P(2,1) \\
&= (1/6)(1/6) + (1/6)(1/6) \\
&= 1/36 + 1/36 \\
&= 2/36
\end{aligned}
$$

a. Complete the probability distribution for the total of two dice in one toss.

X	Probability (*P*)
2	1/36
3	2/36
Total	________

b. What it the probability of a roll of the dice totaling 7? _____

37. The mean number of hours students spent studying for an American History final exam is 12.
The standard deviation is 2.60. How many standard deviations away from the mean are the hours spent by the following students?

a. Kevin, 6 hours _____

b. Gabrielle, 20 hours _____

c. Jean, 10 hours _____

d. Veronica, 15 hours _____

38. The scores for a course presentation have a mean of 65 and a standard deviation of 16. Assuming the distribution is normal:

a. What is the *z*-score of person scoring 75? _____

b. What is the *z*-score of person scoring 59? _____

c. What proportion of the distribution is between 50 and 70? _____

39. The mean number of years employees have worked for the ABA Company is 25 and the standard deviation is 4.3. Assuming a normal distribution:

a. What is the *z*-score of a person employed for 15 years? _____

b. What is the *z*-score of a person employed for 39 years? _____

c. What proportion of people have been employed between 20 and 30 years? _____

d. What proportion of people have been employed longer than 35 years? _____

40. Use these statistics from a normal distribution to answer the following questions.

mean = 110 years
standard deviation = 22.5 years

a. What proportion of the distribution lies above 127? _____

b. What proportion of the distribution lies below 50? _____

c. What proportion of the distribution lies between 75 and 100? _____

d. What proportion of the distribution is less than 60 and more than 80? _____

41. Assuming a normal distribution, consider the following scores:

24	20	15
25	12	18
14	24	16

a. What is the mean? _____

b. What is the standard deviation? _____

c. What proportion of scores are above 15? _____

d. What proportion of scores are less than 20? _____

Jonathan's final grades in four of his classes, as well as the means and standard deviations for each class are listed below. Answer questions 42 – 45, given this data. (Assume the distribution is normal).

COURSE	JONATHAN'S GRADE	MEAN	STANDARD DEVIATION
Mathematics	78	82	2.3
College English III	94	90	4.8
Advanced Spanish	81	79	1.1
Biology	92	85	4.6

42. What is the z score for each of Jonathan's grades?

a. Mathematics _____

b. College English III _____

c. Advanced Spanish _____

d. Biology _____

43. What proportion of Jonathan's classmates scored less than him for each class?

a. Mathematics _____

b. College English III _____

c. Advanced Spanish _____

d. Biology _____

44. In comparison to his classmates, what is Jonathan's strongest subject? _______________

45. In comparison to his classmates, what is Jonathan's weakest subject? _______________

6 Samples and Populations

Summary of Equations

z-score for a sample:

$$z = \frac{\overline{X} - \mu}{\sigma_{\overline{X}}}$$

where $\overline{X}$ = a sample mean
μ = mean of means (the true population mean)
$\sigma_{\overline{X}}$ = standard deviation of the sampling distribution of means

Standard error of the mean:

$$\sigma_{\overline{X}} = \frac{\sigma}{\sqrt{N}}$$

where σ = the population standard deviation
N = sample size

Standard error of the proportion:

$$s_p = \frac{\sqrt{P(1 - P)}}{N}$$

where P = sample proportion
N = total number in sample

Confidence intervals:

A. *Population standard deviation known.*

68% confidence interval = $\overline{X} \pm \sigma_{\overline{X}}$
95% confidence interval = $\overline{X} \pm 1.96\sigma_{\overline{X}}$
99% confidence interval = $\overline{X} \pm 2.58\sigma_{\overline{X}}$

where $\overline{X}$ = the sample mean
$\sigma_{\overline{X}}$ = the standard error of the sample mean

B. *Population standard deviation unknown.*

68% confidence interval = $\overline{X} \pm ts_{\overline{X}}$
95% confidence interval = $\overline{X} \pm ts_{\overline{X}}$
99% confidence interval = $\overline{X} \pm ts_{\overline{X}}$

where $\overline{X}$ = the sample mean
$s_{\overline{X}}$ = unbiased estimate of the standard error of the mean
t = table + value

C. *Proportions.*

68% confidence interval = $P \pm S_P$
95% confidence interval = $P \pm (1.96)\ S_P$
99% confidence interval = $P \pm (2.58)\ S_P$

where P = the sample proportion
S_p = the standard error of the proportion

Unbiased estimate of population variance:

A. *Raw scores*

$$\hat{\sigma}^2 = \frac{\Sigma(X - \overline{X})^2}{N - 1}$$

B. *Sample variance*

$$\hat{\sigma}^2 = S^2 \frac{N}{N - 1}$$

Unbiased estimate of population standard deviation:

A. *Raw scores*

$$\hat{\sigma} = \sqrt{\frac{\Sigma(X - \overline{X})^2}{N - 1}}$$

B. *Sample standard deviation*

$$\hat{\sigma} = s\sqrt{\frac{N}{N - 1}}$$

Unbiased estimate of the standard error of the mean:

$$s_{\overline{X}} = \frac{s}{\sqrt{N - 1}}$$

t ratio:

$$t = \frac{\overline{X} - \mu}{s_{\overline{X}}}$$

where $\overline{X}$ = a sample mean
μ = population mean
$s_{\overline{X}}$ = unbiased estimate of the standard error of the mean

Degrees of freedom:

df = N - 1

α (alpha) level:

α = 1 - level of confidence

Multiple-Choice Questions

1. A researcher is interested in the academic abilities of young children who attend private versus public schools in a large U.S. city. Since it is impractical to locate and test all children, a subset of 400 are selected and given various aptitude tests. This technique is known as

   a. decision making.
   b. sampling error.
   c. sampling.
   d. operationalizing.

2. To be most confident in generalizing from a sample to a population, one should obtain a

   a. quota sample.
   b. hypothesis.
   c. standard error of the mean.
   d. random sample.

3. Researchers always have the opportunity to research an entire population.

   a. True
   b. False

4. A sample size may be larger than the population size in certain cases.

   a. True
   b. False

5. Random sampling methods allow a researcher to

   a. confidently estimate the amount of sampling error to be expected in a given sample.
   b. draw a sample whose characteristics are representative of the population from which it was selected.
   c. give every member of a population some known non-zero probability of being selected into the sample.
   d. all of the above are true of random sampling methods.

6. Elizabeth has decided to sample high school students from a list of the entire population by picking one name and then selecting every 7th name. This is an example of what type of sampling.

   a. stratified sampling
   b. simple random sampling
   c. systematic sampling
   d. cluster sampling

7. Kevin wants to sample the employees at a local bank to conduct his research. He has decided to divide the employees into subgroups based on job category, choose samples from those subgroups and join them together as his final sample. What type of sampling is this an example of?

a. stratified sampling
b. simple random sampling
c. systematic sampling
d. cluster sampling

8. You are watching your favorite daytime talk show, which today is about transvestite serial killers. Viewers are asked to telephone a "900" number to register their opinion about whether Democrats or Republicans are more effective at protecting American children from such people. This is an example of which type of sampling design?

a. accidental sampling.
b. quota sampling.
c. simple random sampling.
d. purposive sampling.

9. Simple random sampling would be a way to produce a sample which over-represents a population on some particular variable, such as race.

a. True.
b. False.

10. A researcher is studying the relationship between exam scores and IQ's of college students. He randomly selects 50 colleges from around the nation of Canada, and from each of these colleges he randomly chooses 10 students. Five different exams are randomly selected from each of the student's total set of exams from the past academic year, and each student takes an IQ test. When the study is published, it receives high marks from other researchers for the care with which it was conducted. Which of the following statements is TRUE?

a. The researcher cannot confidently generalize from his sample to the population of all Canadian college students.
b. There is some sampling error in the researcher's sample.
c. The sample mean and variance of exam scores is exactly the same as the mean and variance of all Canadian college students.
d. There is no way to draw an adequately representative sample from the population of all college students.

11. Dr. Munro conducts a study in which kindergarten children exhibit 5.2 aggressive acts per 6 hours of school. Dr. Hernandez repeats this study (using identical methods), and finds that kindergarten children exhibit 6.4 aggressive acts per 6 hours of school. Which of the following statements is TRUE?

   a. Dr. Munro's sample mean better estimates the population mean of all kindergarten children.
   b. Dr. Hernandez's sample mean better estimates the population mean of all kindergarten children.
   c. The differences in the two results is the result of sampling error.
   d. The true population mean must be between 5.2 and 6.4.

12. The distribution of sample means

   a. can only be normal if the distribution of raw scores is normal.
   b. can only be normal if the sample size is greater than 30.
   c. approximates the normal curve, as long as the sample size is reasonably large (30+).
   d. can never be normal.

13. A sample mean is usually the same as a population mean.

   a. True
   b. False

14. A weight-loss clinic claims that members lose an average of 10 pounds in the first two weeks after joining. A statistician decided to test this claim, and randomly sampled 50 new members. After two weeks, the mean weight lost by this sample was 8.5 pounds, with a standard deviation of 1.8 pounds. Based on this sample, the statistician would probably conclude that

   a. he needed to repeat the experiment.
   b. the clinic's claim is probably valid, since obtaining a sample mean of 8.5 when the true population is 10 would happen about 20% of the time.
   c. the clinic's claim is probably false, since obtaining a sample mean of 8.5 when the true population is 10 would happen about 2% of the time.
   d. the clinic's claim is probably valid, since obtaining a sample mean of 8.5 when the true population is 10 would happen about 30% of the time.

15. As the sample size increases, the standard error of the mean will

   a. increase.
   b. decrease.
   c. be constant.
   d. none of the above.

16. From a single population, which of the following distributions has greater variability among its scores?

a. The population distribution.
b. The sampling distribution of means.
c. Neither; both the population distribution and its distribution of sample means have the same variability.
d. Too little information is given to answer this question.

17. A sociobiologist suspects that the circumference of a person's head affects the likelihood of their graduating from high school. She draws 100 samples, each of 15 people, and for each sample calculates a mean head circumference. She then plots the frequencies with which she obtained each mean height. She then repeats this exercise with 100 additional samples drawn from the same population, except that in this second set each sample contains 100 people. Which of the two sets of sample means will have the greatest variability?

a. Too little information is given to answer this question.
b. The distribution based on 100 samples of 100 people.
c. The distribution based on 100 samples of 15 people.
d. They will be the same, since both were based on the same population and on 100 samples.

18. A researcher samples 50 students and calculates a mean GPA of 2.8. He claims with 90% confidence that the mean GPA of the whole university student body is 2.8. This means that

a. there are 5 chances out of 100 that the population GPA is not 2.8.
b. there are 95 chances out of 100 that the population GPA is 2.8.
c. there are 10 chances out of 100 that the population GPA is not 2.8.
d. there are 68 chances out of 100 that the population GPA is 2.8.

19. To calculate a confidence interval, you need

a. the sample mean
b. the population mean
c. the population variance
d. the standard error.
e. both a and c.
f. both a and d.

20. Which of the following statements about confidence intervals is FALSE?

a. As the confidence interval increases from 68% to 95%, the level of confidence increases.
b. The 68% confidence interval has a wider range than does the 95% confidence interval.
c. Confidence intervals can be constructed for any level of probability.
d. The 72% confidence interval has a narrower range than does the 99% confidence

interval.

21. The greater the level of confidence,

  a. the smaller the z-score.
  b. the narrower the confidence interval.
  c. the larger the z-score
  d. the greater the precision

22. An unbiased estimate of population variance is achieved by

  a. dividing the sum of deviations by N.
  b. dividing the sum of deviations by N - 1.
  c. taking the square root of deviations.
  d. using the sample mean.

23. When using an estimate of the standard error of the mean, the sampling distribution of means

  a. follows the normal distribution.
  b. follows the z-distribution.
  c. follows the F distribution.
  d. follows the t distribution.

24. As the degrees of freedom increase, the t distribution

  a. more closely approximates the z distribution.
  b. strays farther from the normal distribution.
  c. more closely approximates the normal distribution.
  d. both a and c.

25. The difference between a t-ratio and a z score is that

  a. the t-ratio is based on population variance.
  b. the z-score is based on an estimate of the standard error based on sample data.
  c. the t-ratio is based on an estimate of the standard error based on sample data.
  d. the t-ratio better approximates a normal curve.

26. Which of the following statements are TRUE regarding estimating proportions?

  a. proportions, unlike means, do not have a sampling distributions.
  b. The sampling distribution of a proportion is skewed and bimodal.
  c. Unlike the t distribution, only the population proportion is unknown when estimating proportions.
  d. The standard error of the proportion is found by dividing the sample proportion by the total number of cases in the sample.

Exercises

27. The following is a list of registered members of a political party. Each member is given a membership number, which preceded his or her name.

01	Shirley Munro	26	Hoijin Wang
02	Robert Roy	27	Karen Stephan
03	Flora Alexandrou	28	Clayton Ham
04	Jasmine Hobeheidar	29	Dimitris Chaos
05	John O'Connor	30	David Kincheloe
06	Tim Black	31	Eiji Yamamoto
07	Bhavani Sitaraman	32	Carole Hanaway
08	Roberto Sanchez	33	Rebecca Spector
09	Dan Millman	34	Amy Rader
10	Joanna Heliotis	35	Mark Whelan
11	Louis Taylor	36	Barbara Heir
12	Daniel Robert	37	Tom Hannon
13	Leueen Malloy	38	Larry Dameron
14	Carolyn Magnan	39	Stephen McGee
15	Elizabeth Williams	40	Mary Ann Somerville
16	Quint Thurman	41	Tammy Moskowitz
17	Muhammad Ali	42	Alan Jacobs
18	Marion Black	43	Rachel Friedman
19	Steve Houldsworth	44	Don Sampson
20	Lisa Lorant	45	Marnie Thompson
21	Barbara Hartwell	46	Edward Stephan
22	Despina Keivon	47	Bill Mahoney
23	Emily Bonita	48	James Inverarity
24	Michael Lewis	49	Jon Simpson
25	Gary Malaney	50	Emma Marika

A pollster wants to interview a sample of 15 people from this group. Use Table H from the text and randomly select 15 persons, using their membership number.

28. In a sample the size of 20 and a population standard deviation of 3.15, what is the standard error of the mean? _____

29. In a sample the size of 300 and a standard error of the mean of 3.42, what is the population standard deviation? _____

30. In a sample the size of 196 and a standard error of the mean of 1.97, what is the population standard deviation? _____

31. Dr. Oyebanjo has measured the heights of 27 male mine workers and found the mean height to be 65 inches. These 27 were randomly selected from around the world from a total population of 31,239 male mine workers. The worldwide mean height of adult males is 70 inches with a standard deviation of 5.4 inches.

a. What is the standard error? _____

b. Determine the 68% confidence interval for the mean height of mine workers. _____

c. Determine the 95% confidence interval for the mean height of mine workers. _____

d. Determine the 99% confidence interval for the mean height of mine workers. _____

32. A researcher took a random sample of 40 girl scouts within the city of Seattle and found that the mean number of boxes of cookies sold by each girl was 32.7. The state administrative office reports that girl scouts in Seattle sell a mean of 45 boxes of cookies, with a standard deviation of 18 boxes.

a. What is the standard error? _____

b. Determine the 95% confidence interval for the mean number of boxes sold. _____

c. The researcher hands in a report to his boss, but forgets to report the level of confidence with which he concludes that the true mean number of boxes of cookies sold is between 28.68 and 36.72. What confidence level was the researcher using?

33. Compute the 95% confidence interval for the following:

$\overline{X} = 437$ $s = 180.34$ $N = 85$

____

34. Use the statistics in Question #32 to compute the 99% confidence interval.

____

35. A sample of 61 business people was randomly selected and given a leadership test, which measures motivation and "internal strength." The mean score was 83.4, with a standard deviation of 6.7.

a. Estimate the standard error of the mean. _____

b. Calculate the degrees of freedom. _____

c. Determine the value of t from table B. _____

d. Construct a 95% confidence interval for the mean score on this test, using the estimated standard error of the mean. _____

35. Assume that for question #34, 320 people had been sampled.

a. What are the degrees of freedom? _____

b. What is the value of t, given a 95% confidence level? _____

For questions 36 – 38, consider the following data. A researcher was interested in researching the number of times people laugh out loud during a viewing of a new comedy show. The researcher observed 50 people as they individually watched this show and recorded each time they laughed out loud. The data is as follows:

X	f
7	6
6	3
5	11
4	13
3	6
2	7
1	3
0	1

36. Calculate the following

a. The mean. _____

b. The standard error of the mean. _____

37. Construct a 95% confidence interval for the mean number of times people laughed out loud as they watched the comedy.

____

38. Construct a 99% confidence interval for the mean number of times people laughed out loud as they watched the comedy.

____

37. A sample of 60 individuals were asked whether they believed that juveniles charged with serious violent offenses should be tried as adults. They were asked to rate their support of trying juveniles as adults on a scale from "0% support" to "100% support". The results are presented below:

ID	Score	ID	Score	ID	Score	ID	Score
01	65	16	83	31	90	46	43
02	56	17	75	32	84	47	86
03	92	18	91	33	76	48	79
04	78	19	81	34	80	49	85
05	80	20	95	35	92	50	58
06	72	21	74	36	79	51	90
07	69	22	70	37	76	52	77
08	81	23	91	38	96	53	88
09	78	24	79	39	61	54	70
10	80	25	83	40	85	55	89
11	86	26	90	41	91	56	92
12	75	27	76	42	77	57	78
13	50	28	82	43	88	58	89
14	92	29	90	44	79	59	67
15	81	30	87	45	83	60	89

a. Take a random sample of 5 scores and compute the mean and standard deviation.

b. Take a random sample of 20 scores and compute the mean and standard deviation.

c. For your answers in parts a and b above, calculate the estimated standard error of the mean and construct a 95% confidence interval for each mean rating score.

_____ (a) _____ (b)

d. Which confidence interval is narrower? _____ Why? ______________________

38. Recently, the Senate voted unanimously to approve a 6% pay raise for all government employees. Three thousand registered voters (selected randomly) were asked whether they agreed with the pay raise, and 82% were found to disagree with it.

a. Calculate the standard error of the proportion. _____

b. Construct a 95% confidence interval for π. _____

39. Two hundred residents of a large city were selected through random sampling and asked whether they thought marijuana should be legalized; 67% said that it should.

a. Calculate the standard error of the proportion. _____

b. Construct a 95% confidence interval for π. _____

c. In words, what does the confidence interval calculated in part b mean?

____

d. Suppose that the state legislature would vote to legalize marijuana if at least 75% of the state's residents approved of its legalization. As an expert social scientist, would you advise the legislature to legalize marijuana, given the confidence interval in part b? _____

7 Testing Differences Between Means

Summary of Equations

Standard Scores:

A. *Raw Scores*

$$z = \frac{X - \mu}{\sigma}$$

where X = raw score
μ = mean of the distribution of raw scores
σ = standard deviation of the distribution of raw scores

B. *Mean Scores*

$$z = \frac{\overline{X} - \mu}{\sigma_{\overline{X}}}$$

where $\overline{X}$ = sample mean
μ = population mean (mean of means)
$\sigma_{\overline{X}}$ = standard error of the mean (standard deviation of the distribution of means)

C. *Sample Mean Differences*

$$z = \frac{(\overline{X}_1 - \overline{X}_2) - 0}{\sigma_{\overline{X}_1 - \overline{X}_2}}$$

or

$$z = \frac{\overline{X}_1 - \overline{X}_2}{\sigma_{\overline{X}_1 - \overline{X}_2}}$$

where $\overline{X_1}$ = mean of the first sample
$\overline{X_2}$ = mean of the second sample
0 = zero, the value of the mean of the sampling distribution of differences between means (we assume that $\mu_1 - \mu_2 = 0$)
$\sigma_{\overline{X_1}-\overline{X_2}}$ = standard deviation of the sampling distribution of mean differences

D. *Proportions*

$$z = \frac{P_1 - P_2}{s_{P_1-P_2}}$$

where P_1 = proportion of the first sample
P_2 = proportion of the second sample
$s_{P_1-P_2}$ = standard error of the difference in proportions

Standard error of the difference between means:

$$s_{\overline{X_1}-\overline{X_2}} = \sqrt{\left(\frac{N_1 s_1^2 + N_2 s_2^2}{N_1 + N_2 - 2}\right)\left(\frac{N_1 + N_2}{N_1 N_2}\right)}$$

where s_1^2 = variance of the first sample
s_2^2 = variance of the second sample

(variances: $s_1^2 = \frac{\Sigma x_1^2}{N_1} - \overline{X}_1^2$ $\quad s_1^2 = \frac{\Sigma x_2^2}{N_2} - \overline{X}_2^2$)

***t* ratio**:

$$t = \frac{\overline{X_1} - \overline{X_2}}{2_{\overline{X_1}-\overline{X_2}}}$$

where $\overline{X_1}$ = mean of group 1
$\overline{X_2}$ = mean of group 2
$s_{\overline{X_1}-\overline{X_2}}$ = standard error of difference between means

Standard Deviation of the Distribution of the Same Sample Measured Twice:

$$s_D = \sqrt{\frac{\Sigma D^2}{N} - (\overline{X}_1 - \overline{X}_2)^2}$$

where s_D = standard deviation of the distribution of the before-after difference scores

D = "after" raw score subtracted from "before" raw score

N = number of cases or respondents in sample

Standard Error of the Difference in Proportions:

$$s_{P_1 - P_2} = \sqrt{P^*(1 - P^*)\left(\frac{N_1 + N_2}{N_1 N_2}\right)}$$

where P* = pooled or combined sample proportion

N_1 = number of cases or respondents in first sample

N_2 = number of cases or respondents in second sample

(pooled sample proportion: $P^* = \frac{N_1 P_1 + N_2 P_2}{N_1 + N_2}$)

Degrees of Freedom:

$df = N_1 + N_2 - 2$ **(independent groups)**

Multiple-Choice Questions

1. Social researchers usually

    a. use a single sample to estimate a population mean.
    b. describe a single sample.
    c. describe a population.
    d. test hypotheses about differences between two or more groups.
    e. test hypotheses about single samples.

2. Which of the following is an example of a null hypothesis?

    a. Attendance at school has a greater positive effect on grades for males than for females.
    b. AIDS prevention campaigns are more effective for individuals over 18 years of age.
    c. Hispanics have a greater risk of congestive heart failure than do Asians.
    d. People from different socioeconomic groups participate in community recycling programs at the same rate.

3. Which of the following statements about the null hypothesis is FALSE?

    a. The null hypothesis states that two (or more) samples have been drawn from the same population.
    b. Acceptance of the null hypothesis means that there is no significant difference between sample means.
    c. Rejecting the null hypothesis means accepting the research hypothesis.
    e. The null hypothesis is usually stated with the hope of rejecting it.

4. Which of the following is an example of a research hypothesis?

    a. Attendance at school has the same effect on grades for males and females.
    b. AIDS prevention campaigns are more effective for individuals over 18 years of age.
    c. Hispanics have the same risk of congestive heart failure as do Asians.
    d. People from different socioeconomic groups participate in community recycling programs at the same rate.

5. The sampling distribution of differences between means

    a. approximates a normal curve only for large samples.
    b. assumes that all sample pairs differ only because of sampling errors.
    c. never involves negative numbers.
    d. has a mean greater than 0.

6. A researcher who rejects the null hypothesis

   a. concludes that the difference between sampling means has a very small probability of lying close to 0 in the sampling distribution of differences between means.
   b. concludes that the difference between sampling means has a very large probability of lying close to 0 in the sampling distribution of differences between means.
   c. must then test the research hypothesis.
   d. only does so after sampling many, many times.

7. When it is said that a difference between group means is statistically significance at a level of .05, it means that

   a. the observed differences between the sample means would be likely to occur due to sampling error alone more than 95 percent of the time.
   b. the observed differences between the sample means would be likely to occur due to sampling error alone no more than 5 percent of the time.
   c. there is no chance that sampling error could produce mean differences that large.
   e. there is no chance that anything but sampling error could have produced mean differences of that magnitude.

8. A researcher conducts an experiment on alcohol and aggression, with two groups of 40 randomly assigned subjects. One group consumes a drink containing alcohol while the other consumes a non-alcoholic beverage, and none of the subjects know which type of drink they are given. Subjects are then placed in a frustrating situation (an impossible task with blocks), and aggressive responses are recorded. The alcohol consumers produced a mean of 12 aggressive acts, while the others produced a mean of 3 aggressive acts. Translating this difference into a standard score results in a value of 1.8. What is the probability that a mean difference of 9 or more can happen strictly due to sampling error?

   a. 46.41%
   b. 3.59%
   c. .0359
   d. .0718

9. What does the probability in question #8 mean?

   a. A mean difference of 9 or more is likely to occur due to sampling error 46 times in every 100 mean differences.
   b. A mean difference of 9 or more is likely to occur due to sampling error 4 times in every 100 mean differences.
   c. A mean difference of 9 or more is likely to occur due to sampling error 7 times in every 100 mean differences.
   d. The researcher should reject the null hypothesis.
   e. The researcher should accept the null hypothesis.

10. For a t-test, you would reject the null hypothesis when

a. the obtained t value meets or exceeds the critical t value.
b. the obtained t value is less than or equal to the critical t value.
c. you have two samples of less than 30.
d. the α level is .05 or less.

11. Statistical significance refers to

a. a cut-off point.
b. establishing a significance level.
c. rejection regions.
d. all of the above.

12. The alpha level is the

a. level of probability at which the null hypothesis can be accepted with confidence.
b. level of probability at which the research hypothesis can be rejected with confidence.
c. level of probability at which the null hypothesis can be rejected with confidence.
d. Type II error.

13. Type I error is

a. rejecting the null hypothesis when it should have been accepted.
b. accepting the null hypothesis when it should have been rejected.
c. rejecting the null hypothesis when it should have been rejected.
d. accepting the null hypothesis when it should have been accepted.

14. Type II error is

a. rejecting the null hypothesis when it should have been accepted.
b. accepting the null hypothesis when it should have been rejected.
c. rejecting the null hypothesis when it should have been rejected.
d. accepting the null hypothesis when it should have been accepted.

15. For a two-tailed test, the following data applies:

Obtained t = 2.354 df = 12

At what level of significance is our obtained t significant?

a. .02
b. .10
c. .01
d. .001

16. For a two-tailed test, the following data applies:

Obtained t = 2.242 df = 37

At what level of significance is our obtained t significant?

a. .05
b. .02
c. .01
d. .001
e. none of the above

17. For a one-tailed test, the following data applies:

$N_1 = 12$ $N_2 = 12$ Obtained t = 1.074

At what level of significance is our obtained t significant?

a. .025
b. .01
c. .10
d. .05
e. none of the above

18. For a one-tailed test, the following data applies:

$N_1 = 3$ $N_2 = 5$ Obtained t = 1.934

At what level of significance is our obtained t significant?

a. .025
b. .01
c. .10
d. .05
e. none of the above

19. A large difference between two sample means could occur if
a. one mean is very small.
b. one mean is very large.
c. one is moderately small while the other is large.
d. all of the above.

20. When a Type I error becomes less likely, it means that a Type II error simultaneously becomes

a. more likely.
b. less likely.

21. A researcher is interested in the influence of viewing televised violence on subsequent aggression, hypothesizing that subjects viewing violent material exhibit heightened feelings of aggression. Ten subjects are randomly selected and given a test designed to measure aggressive feelings. The subjects view a television drama determined to be violent, and they again are administered the aggression test. Suppose that the critical region is bounded by -1.96 and 1.96, and the researcher calculates a t value of 2.02. The researcher should

a. accept the null hypothesis.
b. reject the null hypothesis.
c. reject the research hypothesis.
d. not enough information to determine.

22. Which of the following is NOT necessary for using a t ratio as a test of significance?

a. A normal distribution of the population.
b. Random samples drawn from the population.
c. Interval or ratio data.
d. Equal population variances, when dealing with independent samples.
e. Samples large enough (i.e., over 30) to ensure a normal distribution.
f. Comparing one independent sample to another, or a single sample measured twice.

23. The phrase "accept the null hypothesis" means

a. that the population means are equal.
b. that the population means are believed to be equal.
c. that the research hypothesis has been proven to be correct.
d. that we are unable to reject the null hypothesis.

Exercises

24. You would use a two-tailed t-test when ____________, and you would use a one-tailed t-test when __________.

25. A sampling distribution of differences between means is __________.

26. Consider the following data:

Republicans	**Democrats**
N = 11	N = 16
s = 12	s = 11
mean = 67	mean = 70

a. Compute the standard error of the difference between means.

b. Compute the t value.

c. Is the mean difference significant with an alpha of .01?

d. Is the mean difference significant with an alpha of .05?

27. Students at Griffin High School and Windsor High School have taken a standardized test. A sample of 48 students that took the test at Griffin have a mean of 440 and a standard deviation of 110. A sample of 37 students that took the test from Windsor have a mean of 425 and a standard deviation of 65.

a. Compute the standard error of the difference between means.

b. Compute the t value.

c. Is the mean difference significant with an alpha of .05?

28. A researcher was interested in the effect of stress on cigarette smoking. Sixteen subjects who smoke cigarettes were selected and were asked to compose an essay on the meaning of life. Eight of them composed their essays while seated in an uncomfortable chair, and were given very negative feedback on their essay quality (high stress condition). The other 8 composed their essays in comfort and were given positive feedback (low stress condition). All subjects were then taken to a room where smoking was allowed (and cigarettes were supplied for those who didn't have any) and were asked to wait for an hour before continuing the experiment. The number of cigarettes smoked by people in each group was recorded and is presented below. The null hypothesis is that there will be no difference in cigarette consumption between people in the two different stress conditions.

High Stress	Low Stress
2	1
2	3
4	2
5	2
3	4
1	2
4	1
3	1

a. What is the mean number of cigarettes smoked by the high stress sample? _____

For the low stress sample? _____

b. What is the standard deviation for each sample? _____

c. Calculate the standard error of the difference between means. _____

d. Calculate the *z* score or *t* ratio, whichever is appropriate. _____

e. Determine the degrees of freedom. _____

f. Assume an alpha of .05. What is the critical score or ratio? _____

g. What research decision should be made? _____

h. Why did you use the score or ratio that you did?

____

29. A researcher is interested in testing the difference between two sample means. The first sample has 36 people in it and the second sample has 40 people in it. Identify the degrees of freedom the researcher would use in the table to find the critical t value. __________

30. Given the following information for a two-tailed test, is the obtained t significant? __________

$N_1 = 20$ $N_2 = 15$ Obtained $t = 2.420$ $\alpha = .05$

31. Dr. Williams develops another hypothesis. She feels that after four years of college, a person's self-perception should be different. Dr. Williams randomly selects 6 students just starting college and administers the OSPT. Four years later, she administers the OSPT to these same students. The results are:

Year 1	Year 4
13	18
10	15
15	12
11	20
8	21
12	19

a. What is the null hypothesis? __________

b. What is D (the difference score)? _____

c. What is the standard deviation of the distribution of difference scores? _____

d. What is the standard error of the mean difference? _____

e. Assume an alpha of .10. What is the critical *t* value? _____

f. What is Dr. Williams' research decision? __________

32. A business consulting firm was interested in determining the effectiveness of a training course for their motivational speakers. It is this firm's policy to have clients evaluate their speaker's performance. A simple random sample of 10 different speakers was obtained, and the number of positive client reviews for their most recent speech was obtained. Each speaker then went through a two-week training course, and the number of positive reviews for their subsequent speech was recorded. The data are presented below.

Speaker	Before Course	After Course
A	30	36
B	12	20
C	21	18
D	42	47
E	19	22
F	28	27
G	31	32
H	33	35
I	39	50
J	24	27

a. What is the research hypothesis? __________

b. What is the standard deviation of the distribution of difference scores? _____

c. What is the standard error of the mean difference? _____

d. Assume an alpha of .01. What is the critical t value? _____

e. What should the consulting firm conclude about their training program?

____

f. Assume an alpha of .05. What should the firm conclude now? __________

33. Given the following information for a two-tailed test, is the obtained t significant?

____

$N_1 = 10$ $N_2 = 9$ Obtained t = 2.568 $\alpha = .01$

34. Determine whether the difference in the means of the two groups is statistically significant.

_____

Group 1	Group 2
N = 18	N = 20
$s^2 = 3$	$s^2 = 2$
= 12	= 9

35. A researcher conducts an experiment to determine if listening to classical music affects reading comprehension. A sample of participants read a passage in silence and was tested for the number of facts they remembered. The sample then read a passage while listening to classical music. They were again tested for the number of facts they remembered. The number of facts remembered before and after are as follows:

Participant	Before Music	After Music
A	10	15
B	8	9
C	9	8
D	5	7
E	7	7
F	8	11
G	10	9
H	6	8

a. Calculate the standard deviation for difference. _____

b. Calculate the standard error of the mean difference. _____

c. Calculate the t ratio. _____

d. Should the researcher accept or reject the null hypothesis at .01? _____

e. Should the researcher accept or reject the null hypothesis at .05? _____

36. A professor is interested in whether a new lecture style improves test scores of his students. Given the following data, test the null hypothesis at the .05 level of significance. Would the professor accept the null hypothesis? __________

Student	Before	After
1	80	85
2	90	90
3	75	70
4	60	75
5	80	95
6	70	80

37. The professor in question #36 was also interested in possible differences between the test scores of sophomores and juniors after listening to the new lecture style. He compile this data:

Sophomores	**Juniors**
70	95
82	75
79	74
91	83
	92
	81

a. Calculate the standard error of the difference between means. _____

b. Calculate the t value. _____

c. Is there a significant difference at the .05 alpha level? _____

d. Is there a significant difference at the .01 alpha level? _____

38. Determine whether the difference in the proportions of people in each group who identify themselves as having alcohol abuse problems is statistically significant (at $\alpha = .05$).

	Police Officers	Dentists
sample size	N = 30	N = 24
# who abuse alcohol	f = 6	f = 8

8 Analysis of Variance

Summary of Equations

Total Sum of Squares:

A. *Definitional formula*

$$SS_{total} = \sum(X - \overline{X}_{total})^2$$

where X = any raw score
$\overline{X}_{total}$ = total mean for all groups combined

B *Computational formula*

$$SS_{total} = \Sigma X^2_{\ total} - N_{total}\overline{X}^2_{\ total}$$

where ΣX^2 = all the scores squared and then summed
N_{total} = total number of subjects in all groups combined
$\overline{X}_{total}$ = total mean of all groups combined

Within Groups Sum of Squares:

A. *Definitional formula*

$$SS_{within} = \Sigma(X - \overline{X}_{group})^2$$

where X = any raw score
$\overline{X}_{group}$ = mean of the group containing the raw score

B. *Computational formula*

$$SS_{within} = \Sigma X^2_{total} - \Sigma N_{group} \overline{X}^2_{group}$$

where		
ΣX^2	=	all the scores squared and then summed
N_{group}	=	number of subjects in any group
$\overline{X}_{group}$	=	mean of any group

Between Groups Sum of Squares:

A. *Definitional formula*

$$SS_{between} = \Sigma N_{group} (\overline{X}_{group} - \overline{X}_{total})^2$$

where		
N_{group}	=	number of subjects in any group
$\overline{X}_{group}$	=	mean of any group
$\overline{X}_{total}$	=	mean of all groups combined

B. *Computational*

$$SS_{between} = \Sigma N_{group} \overline{X}^2_{group} - N_{total} \overline{X}^2_{total}$$

where		
N_{group}	=	number of subjects in any group
$\overline{X}_{group}$	=	mean of any group
N_{total}	=	total number of subjects in all groups combined
$\overline{X}_{total}$	=	total mean of all groups combined

Mean Squares (Variances):

A. *Between groups*

$$MS_{between} = \frac{SS_{between}}{df_{between}}$$

where		
$MS_{between}$	=	between-groups mean square
$SS_{between}$	=	between-groups sum of squares
$df_{between}$	=	between-groups degrees of freedom

B. *Within groups*

$$MS_{within} = \frac{SS_{within}}{df_{within}}$$

where MS_{within}	=	within-groups mean square
SS_{within}	=	within-groups sum of squares
df_{within}	=	within-groups degrees of freedom

Degrees of Freedom:

A. *Between Groups*

$$df_{between} = k - 1$$

where k = number of groups

B. *Within Groups*

$$df_{within} = N_{total} - k$$

where N_{total}	=	total number of scores in all groups combined
k	=	number of groups

F-Ratio:

$$F = \frac{MS_{between}}{MS_{within}}$$

Tukey's HSD (Honestly Significant Difference):

$$HSD = q\sqrt{\frac{MS_{within}}{N_{group}}}$$

where q	=	a table value at a given level of significance for the maximum number of means being compared
MS_{within}	=	within-groups mean square (obtained from the analysis of variance)
N_{group}	=	number of subjects in each group (assumes the same number in each group). For unequal groups, use the harmonic mean of the groups sizes, which is the k-th root of the product of all the group sizes (where k is the total number of groups).

Multiple-Choice Questions

1. A researcher is testing the effectiveness of a program designed to help juvenile delinquents increase school attendance. He wants to examine three groups: A group of serious offenders (those arrested more than 4 times), a group of moderate offenders (arrested 2 to 4 times), and a group of first-time offenders (arrested only 1 time). If the researcher calculates 3 separate t-tests for the serious versus moderate, serious versus first-time, and moderate versus first-time, he will be

    a. increasing the probability of a Type II error.
    b. reducing the alpha level
    c. reducing sampling error
    d. increasing the probability of a Type I error.

2. If you conducted an experiment with 10 different groups and wanted to perform separate t-tests, you would have to make 45 different comparisons. If you used an alpha level of .10 you would expect to find about _____ of these comparisons to be significant due to sampling error alone.

    a. 5
    b. 1
    c. 10
    d. 23

3. Rather than make 45 comparisons on the 10 groups from question #2, you can use one test which tells you whether a significant differences are present between any two of the 45 pairs of groups compared. This test is known as

    a. a t-test.
    b. analysis of variance.
    c. Tukey's HSD.
    d. F-ratio.

4. The deviation of raw scores from their group mean is known as

    a. within-groups variation.
    b. total variation.
    c. between-groups variation.
    d. sum of squares.

5. The deviation of group means from one another is known as

    a. within-groups variation.
    b. total variation.
    c. between-groups variation.
    d. sum of squares.

6. In an analysis of variance, the larger the between-groups variance estimate is in comparison to the within-groups variance estimate, the more likely it is that

    a. sampling error has occurred.
    b. the samples came from populations with equal means.
    c. a Type I error has been made.
    d. the samples do not come from populations with equal means.

7. The deviation of a sample mean from the total mean would be used in calculating the

a. total sum of squares.
b. between-group sum of squares.
c. within-group sum of squares.
d. within-groups df.

8. The deviation of a single score from the total mean would be used in calculating the

a. total sum of squares.
b. between-group sum of squares.
c. within-group sum of squares.
d. within-groups df.

9. The deviation of a single score from the sample mean would be used in calculating the

a. total sum of squares.
b. between-group sum of squares.
c. within-group sum of squares.
d. within-groups df.

10. Which of the following statements if FALSE?

a. The sum of squares becomes larger as variation in a set of scores increases.
b. The total sum of squares is partitioned in an analysis of variance into a between-groups sum of squares and a within-groups sum of squares.
c. Larger sample sizes will tend to yield a smaller sum of squares.
d. The F ratio indicates nothing about differences between any pair of group means.

11. The F ratio compares

a. total variation to variation between groups.
b. total variation to variation within groups.
c. variation within groups to variation in the total sample.
d. variation between groups to variation within groups.

12. The larger the F ratio, the more likely is

a. a significant result.
b. Type II error.
c. Type I error.
d. failure to reject H_o.

13. Given a within groups mean square of 11 and a between groups mean square of 28, what is the obtained F ratio?

a. 1.54
b. 0.39
c. 2.55
d. 2.05
e. none of the above

14. Given a within groups mean square of 16.27 and a between groups mean square of 69.31, what is the obtained F ratio?

a. 4.31
b. 3.82
c. 2.26
d. 4.27
e. none of the above

15. Given the following data, what is the obtained F ratio?

$SS_{between} = 73.2$ $SS_{within} = 19.9$
$N_{total} = 12$ $k = 4$

a. 9.80
b. 7.99
c. 6.94
d. 9.70
e. none of the above

16. Given the following data, what is the obtained F ratio?

$SS_{between} = 15.7$ $SS_{within} = 37.8$
$N_{total} = 9$ $k = 3$

a. 1.52
b. 1.36
c. 1.46
d. 2.25
e. none of the above

17. Which of the following is NOT a requirement for using the F ratio?

a. The population variances for the different groups should be equal.
b. Extreme sample variances can be ignored.
c. Samples must be randomly selected.
d. The sample characteristic measured is normally distributed in the population.

18. A researcher is studying three different age groups and each group's attitude toward nude sun-bathing. Each group contains 5 people, each of whom is given a questionnaire asking for their attitudes. On a ten-point scale where lower scores indicate more approval of nude sun-bathing, the mean for young people was 5.2, for middle-aged people was 5.6, and for older people was 6.1. Performing the analysis of variance yields an F ratio of 4.21, which is significant at the .05 level. From this F ratio, the researcher can safely conclude that

a. young people feel significantly more favorable toward nude sun bathing than do middle aged or older people.
b. young people feel as positively about nude sun bathing as middle aged people, and both groups feel more positively about it than do older people.
c. none of these groups is significantly different than any of the other groups in their opinions about nude sun bathing.
d. there is a significant difference among the three groups regarding nude sun bathing.

19. Which of the following is TRUE?

a. A researcher performs an analysis of variance and obtains a non-significant F ratio. She may now use Tukey's HSD to determine differences not detected by the analysis of variance.

b. A researcher is conducting an experiment with three groups. Because of computer difficulties, he loses 3 subjects in the first group, 1 subject in the second, and none in the third group, resulting in groups of unequal size. The researcher may still perform an analysis of variance, but cannot use Tukey's HSD.

c. A mean difference of 3.2 is more likely to be significant in a multiple comparison of 3 groups than in a multiple comparison of 8 groups.

20. A researcher is interested in evaluating the scores of males and females on the Eating Disorders Test, designed to measure attitudes about eating and the prevalence of eating disorders. He performs an analysis of variance comparing males and female and calculates an F ratio of 9.0. If he had instead performed a t-test, he would have calculated a t of

a. 9.0
b. 2.0
c. 4.5
d. 3.0

Exercises

21. A researcher is studying the effects of drugs on task performance. Subjects are tested under 3 different conditions. In one condition, subjects are given marijuana. In another condition, they were given alcohol, and in a third condition they were not given any drug. All subjects then performed a simple task of placing differently-shaped blocks into appropriate places on a wooden board. The hypothesis is that the drugs will affect task performance. The results (representing the number of correctly placed blocks) are presented below:

Marijuana	Alcohol	Control
5	2	8
3	3	6
6	0	7
5	2	4
7	1	8
4	4	7
6	2	6

a. What is the mean for the group receiving marijuana? _____

b. What is the mean for the group receiving alcohol? _____

c. What is the mean for the control group? _____

d. What is the total sum of squares? _____

e. What is the within-groups sum of squares? _____

f. What is the between-groups sum of squares? _____

22. Given five groups, what is the between groups degrees of freedom? _____

23. Fill in the blanks in the analysis of variance summary table below.

Source of Variation	df	SS	MS	F
Between Groups	2	62.1		
Within Groups		37.2	6.2	
Total		99.3		

24. Fill in the blanks in the analysis of variance summary table below.

Source of Variation	df	SS	MS	F
Between Groups	3			4
Within Groups	4	14		
Total	7			

25. Consider the following data:

$SS_{between} = 42$ $SS_{within} = 45$
$N_{total} = 20$ $k = 5$

a. At the .05 level of significance, would we need to calculate a Tukey's HSD? Why/Why Not?

____

b. At the .01 level of significance, would we need to calculate a Tukey's HSD? Why/Why Not?

____

26. Consider the following data:

$MS_{between} = 24$ $MS_{within} = 6$
$N_{total} = 12$ $k = 4$

a. At the .05 level of significance, would we need to calculate a Tukey's HSD? Why/Why Not? ____________________

b. At the .01 level of significance, would we need to calculate a Tukey's HSD? Why/Why Not? ____________________

27. A business is interested in comparing three different sales techniques: phone sales, coupons, and direct contact. Twenty-four sales people are selected and randomly assigned to 3 different groups of 8. Each group is trained in one of the three different techniques, and each person's sales were observed during a 1-month follow-up period.

phone sales	coupons	direct contact
13	10	16
15	8	12
10	11	17
12	5	13
16	7	20
14	9	18
13	10	19
12	9	15

a. What is the mean number of sales for each group? _____

b. What is the F ratio? _____

c. Assuming an alpha of .01, what is the critical F ratio? _____

d. What is the research conclusion? _______________

28. Conduct a multiple comparison of means using Tukey's method (using an alpha of .01) for the data in question #27.

a. Construct a table of differences between ordered means.

b. What is the value of q? _____

c. Calculate HSD. _____

d. Which mean difference(s) is/are significant? _____

e. Which mean differences are not significant? _____

29. Given the following data, test the hypothesis that there is no significant difference between group means.

Group 1	Group 2	Group 3
3	4	6
6	9	7
7	8	3
8	5	8
5	3	6

a. What is the research conclusion? _____

b. Should you perform a Tukey's HSD multiple comparison on these means? _____

Why or why not? ___________

30. Based on his own experiences in court, a prosecutor believes that some judges provide more severe penalties than others for people convicted for domestic assault and battery. A sample of five of the most recent domestic assault and battery sentences (in months) handed down by each of three judges are shown below.

Judge # 1	Judge # 2	Judge #3
1	3	1
1	2	5
3	4	2
2	3	1
2	4	1

a. Test the hypothesis that some judges give longer sentences than others for the same offense (at $\alpha = .05$). _____

b. Does Judge #1 give significantly shorter sentences than Judge #2? _____

9 Nonparametric Tests of Significance

Summary of Equations

Chi-Square:

$$X^2 = \Sigma \frac{(f_o - f_e)^2}{f_e}$$

where f_o = observed frequency in any category
f_e = expected frequency in any category

Chi-Square with Yates' correction for small f_e:

$$X^2 = \Sigma \frac{(|f_o - f_e| - .5)^2}{f_e}$$

where f_o = observed frequency in any category
f_e = expected frequency in any category

Degrees of Freedom for Chi-Square:

A. *One-way*

df = k - 1

where k = number of categories in the observed frequency distribution

B. *N-way*

df = (r - 1)(c - 1)

where r = number of rows
c = number of columns

Expected Frequencies:

$$f_e = \frac{\text{(row marginal total)(column marginal total)}}{N}$$

Cross-Tabulation:

A. *Total*

$$\text{tot\%} = (100)\frac{f}{N_{tot}}$$

where f = frequency in any category
N_{tot} = total sample size

B. *Row*

$$\text{row\%} = (100)\frac{f}{N_{row}}$$

where f = frequency in any row
N_{row} = number of cases in that row

C. *Column*

$$\text{col\%} = (100)\frac{f}{N_{col}}$$

where f = frequency in any column
N_{col} = number of cases in that column

Mann-Whitney U Test:

$$U = N_1N_2 + \frac{N_1(N_1 + 1)}{2} - \Sigma R_1$$

(or)

$$U' = N_1N_2 + \frac{N_2(N_2 + 1)}{2} - \Sigma R_2$$

where ΣR_1 = sum of the ranks for the smaller group
ΣR_2 = sum of the ranks for the larger group
N_1 = size of the smaller sample
N_2 = size of the larger sample

Z-Score (Mann-Whitney):

$$z = \frac{\dfrac{U - N_1 N_2}{2}}{\sqrt{\dfrac{N_1 N_2 (N_1 + N_2 + 1)}{12}}}$$

where U = Mann-Whitney test
N_1 = size of the smaller sample
N_2 = size of the larger sample

Kruskal-Wallis H:

$$H = \frac{12}{N(N+1)} \Sigma \left[\frac{(\Sigma R_i)^2}{n} \right] - 3(N+1)$$

where N = total number of cases
n = number of cases in a given sample
ΣR_1 = sum of the ranks for a given sample

Degrees of Freedom for Kruskal-Wallis H:

df = k - 1

where k = number of groups/conditions/categories

Multiple-Choice Questions

1. Analysis of variance, the t ratio, and other tests of significance which require normality in the population and interval or ratio level measurement of the data are known as __________ tests.

   a. parametric
   b. nonparametric
   c. rank
   d. chi-square

2. In general, the power of parametric tests is __________ that of nonparametric tests.

   a. greater than
   b. less than
   c. about equal to
   d. none of the above

3. The set of frequencies obtained in an actual frequency distribution are called the

   a. expected frequencies.
   b. observed frequencies.
   c. acquired frequencies.
   d. marginal percentages.

4. The set of frequencies that are expected to occur under the terms of the null hypothesis are called the

   a. expected frequencies.
   b. observed frequencies.
   c. acquired frequencies.
   d. marginal percentages.

5. In a chi-square test, as the difference between f_o's and f_e's increases,

   a. the likelihood of failing to reject H_o increases.
   b. the critical value of chi-square increases.
   c. the likelihood of rejecting H_o increases.
   d. the critical value of chi-square decreases.

6. A researcher uses a chi-square test to investigate buying strategies among people purchasing CD players. She wants to determine who are the buyers of four different brands of CD player. The null hypothesis for this research would be

   a. in the population of CD player buyers, one brand will be preferred over the other three brands.
   b. that there are real differences in the population regarding CD player preferences.
   c. that 1/4 of the sample will prefer each brand.
   d. that 1/4 of the population will prefer each brand.

7. A researcher uses a two-way chi-square test to investigate the television viewing habits of people in a city with 4 local television stations. The researcher used a random sample composed of men, women, and children. What are the degrees of freedom?

a. 7
b. 6
c. 12
d. 1

8. When testing the differences between groups, the degrees of freedom for the chi-square test are derived by

a. k - 1.
b. (r - 1)(c - 1).
c. r x c.
d. N(k - 1).

9. Which of the following statements about the chi-square test is TRUE?

a. If the expected frequency in any cell is below 10, a chi-square test should not be used.
b. Misleading results could be obtained if the expected frequency in any cell is below 10.
c. Yates' correction is used for the chi-square test when the expected frequency in any cell is greater than 10.
d. If the expected frequency in any cell is below 5, a chi-square test should not be used, even if the problem cell(s) don't have much of an impact on the value of chi-square.

10. Compared to a parametric test, chi-square

a. has more assumptions about the data and underlying distribution.
b. can be used indiscriminantly.
c. requires larger samples.
d. makes few demands on the data and underlying distribution.

11. Which of the following is NOT a requirement for use of the chi-square test?

a. A comparison between two or more independent samples is needed.
b. The sample data is expected to reflect a normal population distribution.
c. Samples must be drawn randomly from the population.
d. The expected cell frequencies should not be too small.

12. A drawback to the median test is that it ignores

a. the number of cases.
b. the number of variables.
c. the rank order of cases.
d. the median(s) of the sample.

13. Which of the following tests are appropriate substitutes for a t test when dealing with ordinal data?

a. chi-square test
b. Mann-Whitney U test
c. median test
d. both b and c

14. If the size of the larger sample is greater than 40 and the size of the smaller sample is less than 20, it is safe to assume that the U statistic is

a. normally distributed.
b. bimodal and leptokurtic.
c. equivalent to a t.
d. equal to 2 * t.

15. Which test should you utilize when comparing three or more samples containing at least six cases each?

a. Mann-Whitney U
b. Kruskal-Wallace
c. chi-square
d. median

16. In a study of violence content in children's TV shows, researchers were interested in discovering whether different cable or network channels differed in the levels of violence contained in their programs. For one week, 30-minute shows were randomly sampled from 4 different channels. These channels were coded A, B, C, and D. By the end of one week, seven different shows had been recorded from each channel. Two independent researchers who knew nothing about the purposes of the study were asked to rate the shows on a scale from little violence (1) to a great deal of violence (2). In order to test the hypothesis that there is a difference among the networks in the violence content of their children's shows, which of the following should be used?

a. Mann-Whitney U
b. Kruskal-Wallis
c. chi-square
d. median test

17. A research analyst asked a sample of 200 business travelers whether they have used electronic ticketing for their most recent business trip. The sample was composed of 38 financial professionals, 26 health professionals, 63 computer professionals, 34 lawyers, and 39 sales professionals. What is the degrees of freedom for a two-way chi-square test?

a. 200
b. 4
c. 5
d. 3

18. The research analyst in question #17 is using an alpha level of .01. Which obtained chi-square would lead him to reject the null hypothesis?

a. 9.483
b. 7.885
c. 14.345
d. 13.207

19. The research analyst in question #17, decided to use a two-way chi-square test in investigate the answers of different racial groups for each profession. The computer professionals were composed of 26 Caucasians, 21 Asians, 11 African Americans, and 5 Latinos. What is the degrees of freedom for computer professionals?

a. 63
b. 4
c. 2
d. 3

20. In reference to question #19, the research analyst is using an alpha of .01. Which obtained chi-square would lead him to accept the null hypothesis?

a. 11.344
b. 37.536
c. 11.453
d. 12.093

Exercises

21. An independent firm conducts a taste test of 5 different brands of tortilla chips. The chips are coded A, B, C, D, and E. Two hundred subjects are chosen randomly and asked to try each chip and decide which brand they prefer. The results are summarized below:

Brand	Frequency
A	28
B	48
C	35
D	58
E	31

a. What are the expected frequencies under the null hypothesis that people have no tortilla chip preferences? __________

b. What is the chi-square? _____

c. What are the degrees of freedom associated with the chi-square? _____

d. Assuming an alpha of .05, what is the critical value of chi-square? _____

e. What is the statistical conclusion of the chi-square test? __________

22. The senior class of Beach High School was polled to determine their favorite teacher. The results are as follows:

Teacher	Frequency	Teacher	Frequency
A	14	G	15
B	22	H	23
C	25	I	18
D	26	J	15
E	10	K	27
F	22	L	23

a. What are the expected frequencies under the null hypothesis that the seniors have no significant preference of teachers? __________

b. What is the chi-square? _____

c. What are the degrees of freedom associated with the chi-square? _____

d. Assuming an alpha of .05, what is the critical value of chi-square? _____

e. Assuming an alpha of .01, what is critical value of chi-square? _____

f. What is the statistical conclusion for both levels of alpha?

____

23. A researcher is interested in whether there is a relationship between handedness and beer-drinking preference. A random sample of 35 right-handed and 35 left-handed people is drawn, and their preference for one of two major beer brands is recorded:

Handedness

Beer	Right	Left
Schmoltz	25	18
Blatz	10	17
Total	35	35

a. What is the expected frequency of right-handers who prefer Schmoltz? __________

b. What is the expected frequency of right-handers who prefer Blatz? __________

c. What is the expected frequency of left-handers who prefer Schmoltz? __________

d. What is the expected frequency of left-handers who prefer Blatz? __________

e. What is the chi-square? _____

f. What are the degrees of freedom associated with the chi-square? _____

g. Assuming an alpha of .01, what is the critical value of chi-square? _____

h. If the null hypothesis is that there is equal preference of beer brands among right and left-handed people, what is the statistical conclusion of this chi-square test?

____

24. The Schmoltz and Blatz companies learn of the research described in question #23, and ask the researchers to do something more useful like examining the relationships between certain geographic regions and beer preference. A sample of 53 from the Southeast, 49 from the Mid-atlantic, and 44 from the Northeast was obtained and their beer preferences are as follows:

Region

Beer	Southeast	Mid Atlantic	Northeast
Schmoltz	29	32	18
Blatz	24	17	26

a. What is the chi-square? _____

b. What are the degrees of freedom? _____

c. Assuming an alpha of .05, what is the statistical conclusion?

____

25. A professor investigated the preferences in teaching styles of students at his college. He studied three teaching styles: lecture only; discussion only; and lecture and discussion. He sampled 150 students: 25 freshmen; 42 sophomore; 37 juniors; and 46 seniors. The results are as follows:

Students

Style	Freshmen	Sophomore	Juniors	Seniors
Lecture only	10	14	5	3
Discussion only	9	6	14	15
Lecture & discussion	6	22	18	28

a. What is the chi-square? _____

b. What are the degrees of freedom associated with the chi-square? _____

c. Assuming an alpha of .05, what is the statistical conclusion? _______________

d. Assuming an alpha of .01, what is the statistical conclusion? _______________

26. Professor Phillips is conducting a study on how many times per month men and women, from a particular neighborhood, use ATM machines. She has sampled 12 men and 13 women. The table below shows the data she collected.

Men	**Women**
10	0
5	2
8	10
7	4
4	5
10	7
3	4
9	6
5	9
8	8
9	4
6	6
	6

a. Find the median of both men and women combined. _____

b. How many men fall above the median? _____

c. How many women do not fall above the median? _____

d. Using Yates's correction, what is the obtained chi-square? _____

e. Assuming an alpha of .05, what should Dr. Phillips conclude? _______________

27. Dr. Weber studies the influence of various factors on people's self-reports of drug use. In one study, she randomly selects 20 subjects and assigns them to one of two groups. Each group first hears a 10-minute lecture on drug use, and then every subject is asked to estimate the number of times he or she has used illegal drugs. The difference between the two groups is that in one group, the lecture depicts drug use as a highly dangerous, illegal, and immoral activity, while the other hears a lecture portraying drug use as relatively harmless activity that people sometimes choose to entertain themselves. Will subjects hearing different lectures respond differently when asked to report their own use of drugs? Of the 20 subjects chosen, several had to be excluded from the study for various reasons. Data from the experiment are presented below:

Estimate Number of Times
Using Illegal Drugs

Unfavorable Lecture	Favorable Lecture
40	1
9	12
15	34
0	25
18	0
16	46
7	29
3	
1	

a. Rank order all the scores in both groups combined.

b. What is the value of U? _____ What is the value of U'? _____

c. Assuming an alpha of .05, what is the critical value of U? _____

d. What is the null hypothesis? __________

e. What is the statistical conclusion? __________

28. Americans for Decency in Film, a citizen's lobbying group, commissions a study to examine whether motion picture ratings are associated with the number of violent acts in movies. Dr. A.B. Anderson randomly selects 6 movies from each of 6 categories: XXX, NC-17, R, PG-13, PG, and G. A team of research assistants is trained to determine when a movie scene contains violence, views the films, and records the number of scenes in each movie containing violence.

G	PG	PG-13	R	NC-17	XXX
1	5	2	10	12	0
0	2	8	15	6	2
2	3	4	4	1	0
0	1	3	8	4	1
1	4	7	11	8	1
0	3	6	9	2	0

a. Rank order total group of scores.

b. What is the sum of ranks for movies with an R rating? _____

c. What is the value of the Kruskal-Wallis H? _____

d. What are the degrees of freedom associated with the H test? _____

e. Assuming an alpha of .05, what is the critical value of chi-square? _____

f. The null hypothesis is that there are no differences among movies of different ratings in their violent content. What is the statistical conclusion? __________

29. Dr. Ali hypothesizes that men and women are treated differently with regard to repair service. He devised a study where men and women at different times took a single VCR into various repair shops, complained of an identical problem described in exactly the same way, and asked that it be looked at. The price charged (in dollars) for each repair is shown below, broken down by gender of the "customer":

Men	Women
240	120
60	300
100	280
140	160
170	205
150	165
175	210
90	265
100	80
145	190
120	255
110	140
	200

a. What is the median of the two samples combined? __________

b. How many women fall above the median (were charged with more than the median price)? _____ How many men fall below the median (were charged less than the median price)? _____

c. What is the value of chi-square? _____

d. Assuming an alpha of .05, what is the critical value of chi-square? _____

e. What is the null hypothesis? __________

f. What is the statistical conclusion? __________

10 Correlation

Summary of Equations

Pearson's Correlation Coefficient:

A. *Definitional Formula:*

$$r = \frac{\Sigma(X - \overline{X})(Y - \overline{Y})}{\sqrt{\Sigma(X - X)^2 (Y - Y)^2}}$$

where X = any raw score within the X variable
$\overline{X}$ = mean of the X variable
Y = any raw score within the Y variable
$\overline{Y}$ = mean of the Y variable

B. *Computational Formula:*

$$r = \frac{\Sigma XY - N\overline{X}\overline{Y}}{\sqrt{(\Sigma X^2 - NX^2)(\Sigma Y^2 - NY^2)}}$$

where X = any raw score within the X variable
N = the number of pairs of scores X and Y
$\overline{X}$ = the mean of the X variable
Y = any raw score within the Y variable
$\overline{Y}$ = the mean of the Y variable

C. *Sum of Products/Squares:*

$$r = \frac{SP}{\sqrt{SS_x SS_y}}$$

where SP = the sum of products
SS_x = the sum of squares for the X variable
SS_y = the sum of squares for the Y variable

Sum of Products:

$$SP = \Sigma XY - N\overline{X}\,\overline{Y}$$

where X = any raw score within the X variable
N = the number of pairs of scores X and Y
$\overline{X}$ = the mean of the X variable
Y = any raw score within the Y variable
$\overline{Y}$ = the mean of the Y variable

Sum of Squares:

A. *X*

$$SS_X = \Sigma X^2 - N\overline{X}^2$$

where X = any raw score within the X variable
$\overline{X}$ = the mean of the X variable
N = the number of pairs of scores X and Y

B. *Y*

$$SS_Y = \Sigma Y^2 - N\overline{Y}^2$$

where Y = any raw score within the Y variable
$\overline{Y}$ = the mean of the Y variable
N = the number of pairs of scores X and Y

t Ratio (Pearson's r):

$$t = \frac{r\sqrt{N-2}}{\sqrt{1-r^2}}$$

where N = the number of pairs of scores X and Y
r = the obtained Pearson's correlation coefficient

Degrees of freedom:

$$df = N - 2$$

where N = the number of pairs of scores X and Y

Partial Correlation Coefficient:

$$r_{XYZ} = \frac{r_{xy} - r_{xz} r_{yz}}{\sqrt{1 - r_{xz}^2}\sqrt{1 - r_{yz}^2}}$$

where			
	r_{XY}	=	the correlation between X and Y
	r_{XZ}	=	the correlation between X and Z
	r_{YZ}	=	the correlation between Y and Z

Multiple-Choice Questions

1. The correlation coefficient is used to

   a. describe the relationship between two nominal variables.
   b. describe he relationship between two interval-level variables.
   c. describe the relationship between one nominal and one interval-level variable.
   d. all of the above.

2. In a graph with the X variable on the horizontal axis and the Y variable on the vertical axis, a correlation is stronger

   a. the more scattered the points are.
   b. when the points form a square.
   c. when the points form a diagonal.
   d. when the points form a circle.

3. Which of the following correlation coefficients indicates the weakest relationship between two sets of scores or variables?

   a. .16
   b. -.29
   c. .53
   d. -1.00

4. Which of the following correlation coefficients indicates the strongest relationship between two sets of scores or variables?

   a. .25
   b. -.86
   c. .53
   d. .80

5. Smoking (the X variable) and heart disease (the Y variable) have a positive correlation. This means that

   a. the more a person smokes, the less the chance of heart disease.
   b. the less a person smokes, the greater the chance of heart disease.
   c. the less a person smokes, the less the chance of heart disease.
   d. smoking causes heart disease.

6. Income (the X variable) and crime within one's neighborhood (the Y variable) have a negative correlation. This means that

   a. as income increases, neighborhood crime increases.
   b. as income increases, neighborhood crime decreases.
   c. as income decreases, neighborhood crime decreases.
   d. increased income causes crime to fluctuate.

7. In a study on entertainment spending in a small U.S. city, it is found that the relationship between age (the X variable) and spending (the Y variable) is positive for people between the ages of 15 and 44 (that is , the older you are the more you spend on entertainment). However, the relationship between age and spending is negative for individuals between the ages of 45 and 80 (the older you are, the less you spend). The relationship between age and entertainment spending across the life span can be described as

a. strong.
b. positive.
c. negative.
d. curvilinear.

8. Pearson's correlation coefficient uses

a. rankings.
b. frequency data.
c. deviations.
d. chi-square values.

9. With a Pearson's r for a sample, one has a measure of the

a. strength of correlation.
b. effect of one variable on another.
c. direction of correlation.
d. both a and b.

10. A researcher thinks that her variables have a curvilinear relationship. Should she use Pearson's r?

a. yes
b. no

11. The correlation coefficient does NOT measure

a. The variation of one variable with another.
b. the direction of the relationship between two variables.
c. the degree of linear relationship between two variables.
d. the effect of one variable on another.

12. A researcher thinks that his variables have a linear relationship. Should he use Pearson's r?

a. yes
b. no

13. Pearson's r requires both X and Y variables to be normally distributed in the population. This requirement is of only minor importance when the sample size

a. is less than 30.
b. is greater than or equal to 30.
c. is greater than or equal to 15.
d. is between 10 and 30.

14. Pearson's correlation coefficient may be distorted by

a. the presence of an outlier.
b. curvilinear relationships between variables.
c. samples of less than 30 cases.
d. all of the above.

15. In the partial correlation coefficient $r_{AB.C}$, the control variable is

a. A
b. B
c. C
d. none of the above

16. In the partial correlation coefficient $r_{AB.C}$, after the third variable is removed, the remaining correlation is between

a. A and B
b. A and C
c. B and C
d. A, B, and C

17. Under which conditions can an obtained r of .3426 lead a researcher to reject the null hypothesis?

a. sample size = 34; $\alpha = .05$
b. sample size = 36; $\alpha = .01$
c. sample size = 38; $\alpha = .05$
d. sample size = 39; $\alpha = .01$

18. Under which conditions can an obtained r of .5350 lead a researcher to accept the null hypothesis?

a. sample size = 14; $\alpha = .05$
b. sample size = 15; $\alpha = .05$
c. sample size = 30; $\alpha = .01$
d. sample size = 25; $\alpha = .01$

15. A researcher collecting data from 91 individuals has obtained an r of .2320. Which of the following statements is/are true?

a. the researcher would accept the null hypothesis at the .05 level
b. the researcher would reject the null hypothesis at the .01 level
c. both statements are true
d. both statements are false

20. A scientist observes the following X, Y pairs: (33,33), (22,22), (8,8), and (24,24). The correlation between X and Y is

a. positive.
b. 0.
c. negative.
d. .5.

21. A scientist observes the following X, Y pairs: (33,11), (42,6), (38,9), and (20,40). The correlation between X and Y is

a. positive.
b. 0.
c. negative.
d. .5.

22. Dr. Hans Tinker of the Netherlands observes that the higher the population of storks per square kilometer, the greater the human birth rate per square kilometer. He publishes an article in ***The Journal for the Advancement of Not-Too-Bright Researchers*** declaring that, consistent with the myth, his evidence proves that storks assist in human reproduction. His research is challenged by another researcher, Dr. L. Seamus, who asserts that more houses in a given area provide more stork nesting sites. He also suggests that where there is a greater density of houses there is a greater density of people, accounting for the high birth rate per square kilometer. What statistic would be most useful to Dr. Seamus in challenging Dr. Tinker's conclusions?

a. t-ratio
b. curvilinear correlation
c. Pearson's correlation
d. partial correlation

Exercises

23. Consider the following data:

X	Y
2	6
6	14
5	12
4	10
1	4

a. Compute the sum of squares for X and Y. _____ _____
b. Compute the sum of products. _____
c. Compute the Pearson's r. _____

24. Consider the following data

$\Sigma XY = 122$ $N = 8$
$\Sigma X^2 = 318$ $\Sigma Y2 = 168$
$X = 5.75$ $Y = 3.75$

a. Compute the sum of squares for X and Y. _____ _____
b. Compute the sum of products. _____
c. Compute the Pearson's r. _____

25. Consider the following data:

X	Y
6	8
9	11
5	8
7	10
9	9

a. Calculate the sum of products. _____

b. Is the correlation between X and Y scores significant at the .01 level? _____

c. Is the correlation between X and Y scores significant at the .05 level? _____

26. Consider the following correlation matrix for age (X), credit card debt (Y), and salary (Z) for a random sample of 50 people between the ages of 18 and 40.

	Age	Debt	Salary
Age	1.00	0.30	0.55
Debt	0.30	1.00	-0.20
Salary	0.55	-0.20	1.00

a. Which correlation(s) is/are significant at the .05 level of significance? __________

b. Which correlation(s) is/are significant at the .01 level of significance? __________

c. What is the partial correlation between age and debt, holding salary constant? _____

d. Is the partial correlation smaller, equal to or greater than the simple correlation between age and debt? __________

27. A researcher is studying the relationship between job performance scores and job satisfaction scores. The data is as follows:

Performance	Satisfaction
37	40
30	25
42	32
29	27
33	37
43	31
27	32
40	45
32	28
27	31

a. Calculate the Pearson's r. _____
b. Identify the critical value of r, assuming an alpha of .05. _____
c. Is the correlation between job performance and job satisfaction scores significant at the .05 level? _____
d. Identify the critical value of r, assuming an alpha of .01. _____
e. Is the correlation between job performance and job satisfaction scores significant at the .01 level? _____

28. Dr. Lawton is conducting a study on the relationship between IQ scores and the results of a self esteem test (measured on a 15-point scale with 15 being very high and 1 being very low). He has collected the following data:

IQ	**Esteem**
88	2
92	5
97	3
102	10
108	9
115	7
117	8
120	10
123	6
126	12
130	11
133	9
137	14
140	13
145	15

a. What is Pearson's r? _____

b. From Table F, what is the critical value of r, assuming an alpha of .05? _____

c. What is the critical value of r, assuming an alpha of .01? _____

d. Is the correlation significant at either level? __________

e. If so, is the correlation positive or negative? ____________________

29. Below is a correlation matrix for a study on intelligence. X1 is a person's age (in years), X2 is the person's education level (in years), X3 is the person's income level (in thousands of dollars), x4 is the person's height (in inches), and X5 is the person's score on an IQ test.

	X_1	X_2	X_3	X_4	X_5
X_1	1.00	-.31	.68	-.05	.26
X_2	-.31	1.00	.98	.12	.57
X_3	.68	.98	1.00	.30	.69
X_4	-.05	.12	.30	1.00	-.24
X_5	.26	.57	.69	-.24	1.00

a. What is the correlation between a person's age and a person's IQ? _____

b. Is the correlation between a person's age and height positive or negative? _____

c. What is the correlation between a person's income and education level? _____

d. What is the correlation between X2 and X5? _____

e. What is the correlation between X3 and X4? _____

30. A researcher was interested in determining the degree of association between IQ (X) and the number of years of continuous employment (Y). She recruited 9 subjects, each of whom took an IQ test and reported the number of years they have been continuously employed. The results were:

Subject	IQ	Years Employed
A	115	6
B	125	5
C	109	4
D	132	3
E	116	8
F	141	2
G	153	1
H	122	2
I	120	7

a. What is Pearson's r? _____

b. From Table F, what is the critical value of r, assuming an alpha of .01? _____

c. Is the correlation between IQ and years of continuous employment significant at the .01 level? _____

d. Is the relationship positive or negative? __________

31. The researcher in question #24 added a third variable, the number of friends a person says they have. The data are:

Subject	IQ (X)	Years Employed (Y)	Number of Friends (Z)
A	115	6	20
B	125	5	15
C	109	4	18
D	132	3	9
E	116	8	12
F	141	2	3
G	153	1	3
H	122	2	4
I	120	7	10

a. What is the partial correlation coefficient of IQ and years of employment, holding the number of friends constant? _____

b. Is the correlation between IQ and years of employment, partialling out the number of friends, significant at the .01 level? _____

c. Compare this answer with your answer to question #30. State the meaning of this comparison. ____

32. As part of a class project, you test the relationship between a male's height (X) and shoe size (Y). Nine men are randomly selected from your class, and their height (in inches) and shoe size are recorded:

X	Y
60	5.0
72	10.0
76	12.0
80	15.0
65	8.5
68	10.0
70	9.5
71	11.0
73	10.5

a. What is Pearson's r? _____

b. From Table F, what is the critical value of r, assuming an alpha of .05? _____

c. Is the correlation between height and show size significant at the .01 level? _____

33. The following was taken from a recent journal article:

	Smoking	Heart Disease	Age
Smoking (X)	1.00	.55	.32
Heart Disease (Y)		1.00	.94
Age (Z)			1.00

a. What is the correlation between smoking and heart disease? _____

b. What is the correlation between smoking and heart disease, controlling for age?

_____

c. What can you say about the effect of age on the relationship between smoking and heart disease? _____

11 Regression Analysis

Summary of Equations

Regression Equation:

$$Y = a + bX + e$$

where			
	Y	=	dependent variable
	X	=	independent variable
	a	=	Y-intercept
	b	=	slope or regression coefficient
	e	=	error or disturbance term

Slope (Regression Coefficient):

A. *Sum of Products/Squares*

$$b = \frac{SP}{SS_X}$$

where			
	SP	=	sum of products
	SS_X	=	sum of squares for the X variable

B. *Deviations*

$$b = \frac{\Sigma(X - \overline{X})(Y - \overline{Y})}{\Sigma(X - \overline{X})^2}$$

where			
	X	=	any raw score within the X variable
	$\overline{X}$	=	mean of the X variable
	Y	=	any raw score within the Y variable
	$\overline{Y}$	=	mean of the Y variable

C. *Raw Scores*

$$b = \frac{\Sigma XY - N\overline{XY}}{\Sigma X^2 - N\overline{X}^2}$$

where
- X = any raw score within the X variable
- $\overline{X}$ = mean of the X variable
- Y = any raw score within the Y variable
- $\overline{Y}$ = mean of the Y variable

Intercept:

$$a = \overline{Y} - b\overline{X}$$

- $\overline{X}$ = mean of the X variable
- $\overline{Y}$ = mean of the Y variable

Predicted Y:

$$\hat{Y} = a + bX$$

where
- a = intercept
- b = slope of the regression line
- X = a given value of X

Error Term (Disturbance Term):

$$e = Y - \hat{Y}$$

where
- Y = observed score
- $\hat{Y}$ = predicted score

Error Sum of Squares (Residual Sum of Squares):

$$SS_{error} = \Sigma e^2 = \Sigma(Y - \hat{Y})^2$$

where
- Y = observed score
- $\hat{Y}$ = predicted score

Total Sum of Squares:

$$SS_{total} = \Sigma(Y - \overline{Y})^2$$

where
- Y = observed score
- $\overline{Y}$ = mean of the Y variable

Regression Sum of Squares (Explained Sum of Squares):

$$SS_{reg} = SS_{total} - SS_{error}$$

Proportionate Reduction in Error (PRE):

$$PRE = \frac{SS_{total} - SS_{error}}{SS_{total}} \quad \text{(coefficient of determination)}$$

Coefficient of Determination:

$$r^2 = \frac{SS_{total} - SS_{error}}{SS_{total}} \quad \text{(proportionate reduction in error)}$$

Coefficient of Non-Determination:

$$1 - r^2 = \frac{SS_{error}}{SS_{total}}$$

Mean Square Regression:

$$MS_{reg} = \frac{SS_{reg}}{df_{reg}}$$

(df_{reg} = k-1, where k = the number of variables)

Mean Square Error:

$$MS_{error} = \frac{MS_{error}}{df_{error}}$$

(df_{error} = N -1, where N = sample size)

F Ratio:

$$F = \frac{MS_{reg}}{MS_{error}}$$

Multiple Coefficient of Determination (Squared Multiple Correlation):

$$R^2 = \frac{r_{YX}^2 + r_{YZ}^2 - 2r_{YX}r_{YZ}r_{XZ}}{1 - r_{XZ}^2}$$

Multiple-Choice Questions

1. Regression analysis is

   a. exactly like correlation.
   b. a way of determining how important one or more independent variables are in predicting a dependent variable.
   c. a procedure designed to show the degree of association between two variables.
   d. none of the above.

2. In the regression equation Y = a + bX, when X is equal to 0, Y is equal to

   a. X.
   b. b.
   c. a.
   d. 1.

3. The slope in a regression equation represents

   a. the value of Y when X is 0.
   b. the base line amount of Y achieved by all subjects in the sample.
   c. the change in Y when X changes by 1 unit.
   d. the change in X when Y changes by 1 unit.

4. Which of the following is FALSE?

   a. Y can be determined by the intercept, the slope, and the score of the subject on X.
   b. The slope in the regression equation represented differences between subjects.
   c. The intercept in the regression equation represents an amount common to all subjects.
   d. Where the regression line crosses the Y-axis determines the intercept.

5. The regression line will NOT pass through which of the following (X,Y) points?

   a. (0, a)
   b. (b, a)
   c. $(\overline{X}, \overline{Y})$
   d. (1, a+b)

6. If there is a negative correlation between X and Y, then the value of b in the regression equation will be
   a. negative.
   b. positive.
   c. 0.
   d. not enough information.

7. The intercept and any predicted Y score ALWAYS have an interpretable meaning in the regression equation.

   a. true
   b. false

8. Suppose that the correlation between X and Y were 1.00. This would mean that

a. very few of the Y values would lie on the regression line.
b. as the value of X increases, the value of Y decreases.
c. all of the Y values would lie on the regression line.
d. as the value of Y increases, the value of X decreases.

9. If a subject's score on the Y variable lies above the regression line, then

a. the subject's error term is positive.
b. the subject's error term is negative.
c. the subject's predicted Y score is the same as his/her actual Y score.
d. X and Y are positively correlated.

10. The amount of variance in the dependent variable explained by the independent variable is

a. the difference between the regression sum of squares and the error sum of squares.
b. the difference between predicting a person's Y score knowing X and predicting a person's Y score without knowing X.
c. the difference between the total sum of squares and the regression sum of squares.
d. both a and c.

11. As the degree of correlation between X and Y increases,

a. the error sum of squares gets larger relative to the total sum of squares.
b. the regression sum of squares gets smaller relative to the total sum of squares.
c. the amount of variance in Y explained by X increases.
d. none of the above.

12. The coefficient of non-determination can be negative when X and Y are highly negatively correlated.

a. true
b. false

13. A study found that smoking (X) accounts for 52% of the variance in heart disease (Y). The coefficient of non-determination is thus

a. 52%
b. 48%
c. 5.2.
d. not enough information

14. Dr. Arneclev was interested in studying the relationship between race (X) and health problems (Y). After collecting his data, Dr. Arneclev found that race accounted for 36% of the variance in health problems. Other research suggested that socio-economic level was also an important factor, so Dr. Arneclev contacted each of his original subjects and recorded their SES. He then calculated a multiple correlation coefficient of .89. How much variance does the addition of SES explain in the number of health problems?

a. 89%
b. 36%
c. 63%
d. 53%

15. Regression is not as effective as experiments in determining cause and effect.

a. true
b. false

16. A researcher takes 10 people and determines their income and their vocabulary. She finds that people's vocabulary (X) explains 73% of the variance in their income (Y). She now knows that

a. people with greater income have more leisure time to build their vocabulary.
b. vocabulary is the only factor affecting a person's income.
c. a higher vocabulary causes a person to have a higher income.
d. vocabulary and income are strongly associated.

17. Considering the following data, what is the slope?

$\Sigma X = 23$	$\bar{X} = 2.875$	$\Sigma XY = 190$
$\Sigma Y = 47$	$\bar{Y} = 5.875$	$\Sigma X^2 = 95$

a. 1.9
b. .96
c. .76
d. 1.83

18. Using the data in question #17, what is the intercept?

a. 3.115
b. 3.69
c. .4125
d. .6138

19. Using the data in question #17, what is the predicted Y for an X of 5?

a. 9.9125
b. 4.899
c. 7.49
d. 7.915

Exercises

20. A regression line is given by the formula below:

$$\hat{Y} = 28.35 + (1.05)X$$

a. What is the predicted score of a person with an X of 21.20?

b. What is the predicted score of a person with an X of 9.63?

21. Consider the following data:

X	Y
4	3
2	5
5	2
4	5
3	2

a. Calculate the slope. __________

b. Calculate the intercept. __________

c. What is the predicted score of a person with an X of 5? _____

d. What is the predicted score of a person with an X of 3? _____

22. Below are data on 12 randomly selected mothers:

Mother	Years of Education	Number of Children
A	6	5
B	12	1
C	14	2
D	2	4
E	8	3
F	12	2
G	20	1
h	16	2
I	10	3
J	9	4
K	7	3
L	12	1

a. Calculate $\sum X^2$, $\sum Y^2$, and $\sum XY$. _____ _____ _____

b. Calculate SS_X, SS_Y, and SP. _____ _____ _____

c. What is the slope of the regression line? _____

d. What is the intercept of the regression line? _____

e. What is the equation for the regression line? __________

f. About how many children would you expect a non-educated mother to have? _____

g. For each year of education, how many fewer children can a mother expect to have?

_____

23. Consider the following data:

X = 4.6	$\Sigma XY = 448.0$	N = 15
Y = 5.53	$\Sigma X^2 = 393.0$	

a. What is the predicted Y for an X of 9? _____

b. What is the predicted Y for an X of 2? _____

24. A researcher has gathered the following data from seven randomly selected employees on the number of years (X) employees have worked in a particular department and their salaries (Y) in thousands. Following is the data:

No. Years (X)	Salaries (Y)
1	18.5
2	20
2	20.25
4	21.5
5	25
7	26
7	27.5

a. What is the slope of the regression line? __________

b. What is the intercept of the regression line? __________

c. What would the predicted salary be for someone who has worked in the department for 6 years? _____

d. What is the prediction error for someone who has worked in the department for 6 years? __________

25. A researcher develops a hypothesis relating shoe size (X) to intelligence (Y). Suppose that he has randomly selected 30 subjects from a population and records both shoe size and IQ. He obtains an SS_{total} of 22,512 and an SS_{reg} of 81.04.

a. Without looking at the raw data, is there any way to know the correlation between show size and intelligence? ____________________

b. How much of the variance in predicting a person's intelligence does knowing his/her shoe size reduce? _____

c. Assuming an alpha of .05, what is the critical F ratio? _____

d. What is the F ratio obtained from this data? _____ Is this significant? _____

26. Dr. Brown is attempting to predict college freshman GPA (Y) when knowing both high school GPA (X) and SAT scores Z). Collecting a random sample of 350 college freshmen from across the country, he finds the following correlations:

	X	Z	Y
X	1.00	.88	.79
Z	.88	1.00	.92
Y	.79	.92	1.00

a. How much of the variance in predicting a person's college freshman GPA do high school GPA and SAT scores explain? _____

b. Suppose that Dr. Brown first performed a simple regression looking at high school GPA (X) and college freshman GPA (Y), obtaining a correlation between the two of .79. How much variance in predicting college freshman GPA does high school GPA explain? _____ Dr. Brown then added information about a person's SAT scores. How much additional variance does SAT scores explain? _____

27. Dr. Matsuda was interested in the relationship between number of publications (X) and personal satisfaction (Y) in social science faculty. Ten social science faculty members were selected from two universities (5 from each) and were administered the Personal Satisfaction Inventory (PSI). The PSI ranges form 1 (low satisfaction) to 10 (high satisfaction). His data are presented below:

Number of Publications (X)	PSI Score (Y)
2	8
10	5
3	1
12	9
8	5
1	5
13	3
7	6
11	4
9	2

a. Draw a scatterplot of the data.

b. Looking at the scatterplot, what can you say about the relationship between number of publications and personal satisfaction?

____

c. What is the slope of the regression line? _____

d. What is the intercept of the regression line? _____

e. Draw the regression line in the scatterplot above.

f. What is the correlation between the number of publications and personal satisfaction? _____

g. If a faculty member has 10 publications, what would be your prediction as to her personal satisfaction level? _____ Would the satisfaction level be any different for a faculty member with 1 publication? _____

12 Nonparametric Measures of Correlation

Summary of Equations

Spearman's Rank-Order Correlation Coefficient:

$$r_s = 1 - \frac{6\Sigma D^2}{N(N^2 - 1)}$$

where D = difference in rank between X and Y variables
N = total number of cases

Goodman and Kruskall's Gamma:

$$G = \frac{N_a - N_i}{N_a + N_i}$$

where N_a = number of agreements
N_i = number of inversions

z-score (Gamma):

$$z = G\sqrt{\frac{|N_a - N_i|}{N(1-G^2)}}$$

where G = calculated Gamma coefficient
N_a = number of agreements
N_i = number of inversions
N = total number of cases

Phi Coefficient:

$$\phi = \sqrt{\frac{X^2}{N}}$$

where X^2 = calculated chi-square value
N = total number of cases

Contingency Coefficient:

$$C = \sqrt{\frac{X^2}{N + X^2}}$$

where X^2 = calculated chi-square value
N = total number of cases

Cramér's V (Contingency Coefficient):

$$V = \sqrt{\frac{X^2}{N(k - 1)}}$$

where N = total number of cases
k = number of rows or columns, whichever is smaller (if the number of rows and columns are equal, either can be used for k)

Lambda Coefficient:

$$\lambda = \frac{F_{iv} - M_{dv}}{N - M_{dv}}$$

where F_{iv} = sum of the largest cell frequencies within each category of the independent variable
M_{dv} = largest marginal total among categories of the dependent variable
N = total number of cases

Proportional Reduction in Error (PRE):

$$PRE = \frac{\text{errors without independent variables} - \text{errors with independent variable}}{\text{errors without independent variable}}$$

Multiple-Choice Questions

1. In Spearman's rank-order correlation coefficient, 'D' represents

   a. the distance of each score from the mean.
   b. the difference in rank between X scores.
   c. the difference in rank between X and Y variables.
   d. the average deviation of each score from the mean.

2. Unlike other measures of association, Spearman's rank-order correlation coefficient

   a. does not require a linear correlation.
   b. requires interval data.
   c. does not depend on random sampling.
   d. none of the above.

3. Dr. Holmes is interested in studying the relationship between delinquency (high, medium, or low) and social class (upper, middle, or lower). The data are based on 3,000 respondents to a survey, and several research assistants arrange the data as a cross-tabulation. The measure of association he would probably use would be

   a. Spearman's correlation coefficient.
   b. Goodman and Kruskall's gamma.
   c. the phi coefficient.
   d. Pearson's correlation coefficient.

4. A gamma coefficient of .32 indicates
   a. a positive correlation based on dominance of inversions.
   b. a negative correlation based on dominance of inversions.
   c. a positive correlation based on dominance of agreements.
   d. a negative correlation based on dominance of agreements.

5. In order to employ a gamma coefficient for cross-tabulations, which of the following is NOT necessary?

   a. Sample cases must have been obtained randomly from a specified population.
   b. Both X and Y variables must be ranked or ordered.
   c. The underlying population distribution is assumed to be normal.

6. Which of the following correlation coefficients requires a linear relationship between the X and Y variables?

   a. Spearman's r_s
   b. the phi coefficient
   c. Goodman and Kruskall's G
   d. none of the above

7. Assuming a constant sample size, the phi coefficient will increase as chi-square

a. decreases.
b. increases.
c. remains constant.
d. not enough information

8. An advantage of the contingency coefficient over the phi coefficient is that

a. the contingency coefficient can be used to compare more than 2 groups.
b. the phi coefficient is limited to nominal data.
c. the contingency coefficient does not require random sampling.
d. the contingency coefficient is not affected by large fluctuations in chi-square.

9. An advantage of Cramer's V over the contingency coefficient is that

a. V can be used with ordinal data.
b. C is influenced by the size of the table on which data is arranged.
c. C can be used with non-square tables (e.g., 4 x 6).
d. C requires that the sample size be greater than 30, while V does not.

10. A λ of .36 means that we have

a. increased the power of our prediction by 36%.
b. decreased the error of our prediction by 36%.
c. increased the error of our prediction by 36%.
d. decreased the error of our prediction by 12.96%.

11. Lambda differs from the contingency coefficient and Cramer's V in that

a. lambda is nonsymmetrical.
b. lambda is bimodal.
c. lambda is asymmetrical.
d. lambda requires ordinal data.

12. The process of looking deeper into the relationship between independent and dependent variables by controlling for another variable or other variables is called

a. elaboration.
b. partialling.
c. elucidation.
d. cross-tabulation.

13. Which of the following values of λ is the strongest?

a. .35
b. .75
c. -.82
d. .05

14. Julian Curtis is researching the relationship between political affiliation and community service participation and has collected data from 100 survery respondents. He has found the total number of agreements among cases to bre 766 and the total number of inversions among cases to be 1390. Identify the Goodman's & Kruskal's gamma coefficient:

a. -3.455
b. 0.55
c. -0.55
d. -0.29

15. The obtained z score for the data provided in question # 7 is:

a. 0.53
b. -0.56
c. 0.65
d. 0.64

Exercises

16. For the following data, find the Spearman's rank-order correlation coefficient (r_S): _____

X	Y
10	7
9	9
8	3
7	6
6	10
5	1
4	4
3	8
2	2
1	5

17. Joseph Phoenix is interested in determining the degree of association between ranking on a school debate team (1 is highest and 10 is lowest) and hours spent preparing for a debate. He has collected the following data:

Student	Team Rank	Hours of Prep
Kenneth	1	10
Janet	2	8
Charles	3	10
Timothy	4	12
William	5	9
Joshua	6	8
Marcus	7	11
Andrea	8	7
Jeffrey	9	5
Kyle	10	6

a. What is the sum of the squared differences? _____

b. What is the obtained Spearman's rank-order correlation coefficient? _____

c. Is there a significant correlation (at both the .01 & .05 levels) between a person's ranking and the number of hours they spend preparing for a debate? _____

18. A random sample of 18 respondents has been surveyed and from the data the sum of the squared differences is 12.

a. What is the obtained rank-order correlation coefficient? _____

b. What is the critical rank-order correlation coeffiecient at the .05 level? _____

c. What is the critical rank-order correlation coeffiecient at the .01 level? _____

d. What is the research conclusion at both levels of significant?

____

19. A random sample of 10 respondents has been surveyed and from the data the sum of the squared differences is 55.

a. What is the obtained rank-order correlation coefficient? _____

b. What is the critical rank-order correlation coeffiecient at the .05 level? _____

c. What is the critical rank-order correlation coeffiecient at the .01 level? _____

d. What is the research conclusion at both levels of significant? _______________

20. Consider the following data in the form of a cross-tabulation about the relationship between status in college (X) and level of extracurricular activity (Y).

Level of Extracurricular	Graduate	Upperclass Undergrad	Lowerclass Undergrad
High	9	20	11
Medium	15	18	14
Low	25	10	16

a. Calculate N_a. __________

b. Calculate N_i. __________

21. Use the data in question #20 to answer the following:

a. What is the gamma coefficient? _____

b. At the .05 level of significance, what is the research conclusion? _______________

c. At the .01 level of significance, what is the research conclusion? _______________

22. Dr. Simpson has just completed an evaluation of a drug treatment program for criminal offenders who are admitted drug users and who have been convicted of drug dealing. His main question was whether prisoners who complete a drug treatment program are later arrested at the same rate as those who were not in the program. Thirty-eight offenders graduated from a prison drug treatment program, while 29 served their time without such treatment. The records of all 67 offenders were reviewed one year after their release from prison, and Dr. Simpson recorded whether each had been arrested during that year:

	Arrested	
Drug Treatment	Yes	No
Yes	30	8
No	17	12
Total	35	35

a. What is the chi-square? _____

b. What is the value of the phi coefficient? _____

c. What are the degrees of freedom associated with the chi-square? _____

d. Assuming an alpha of .05, what is the critical value of chi-square? _____

e. What is the statistical conclusion of the chi-square test regarding the relationship between prison drug treatment and post-release arrest? __________

23. The table below shows a data regarding people's gender and whether they smoke:

Gender	Smokes: Yes	No
Male	11	16
Female	5	18
Total	16	34

a. What is the chi-square? _____

b. What is the value of the phi coefficient? _____

c. What is the statistical conclusion of the chi-square test regarding the relationship between gender and smoking? __________

24. A research firm is studying the relationship between a person's income and his/her membership in a political party. The firm randomly selects 115 people and records each one's personal income (greater than $50,000, between $20,000 and $49,999, or less than $19,999) and political affiliation (Republican, Democrat, Libertarian, and Independent/Other). The data are presented below:

	Political Party			
Income	Republican	Democrat	Independent	Libertarian
>$50,000	20	5	7	2
20,000 - 49,999	11	18	12	5
0 - 19,999	2	23	8	3
Total	33	46	27	10

a. What is the value of chi-square? _____

b. Calculate the phi coefficient or the contingency coefficient, whichever is most appropriate. _____

c. Assuming an alpha of .05, what is the critical value of chi-square? _____

d. Is the relationship between income and political affiliation significant? __________

e. Explain why using the contingency coefficient would be either appropriate or inappropriate. ____

f. Calculate Cramer's V. __________

g. Is the relationship between income and political affiliation significant, based on Cramer's V? __________

h. What is the difference between the measure of association indicated by the contingency coefficient and the measure of association indicated by Cramer's V?

25. Recall question #25 in Chapter 9. The data are presented below:

Beer	Democrat	Republican	Libertarian
	Political Affiliation		
Schmoltz	38	15	8
Blatz	19	31	9
Loggerstein	16	12	18
Total	73	58	35

a. The value of chi-square for this problem is 25.45. What is the value of the contingency coefficient? _____ Is this value significant at an alpha of .05? _____

b What is the value of Cramer's V? _____ Is this value significant at $\alpha = .05$? _____

c. Suppose that the N in this problem was 40. Without performing the calculations, what would this do to the value of C? _________________ What would this new N do to the value of V? _________________ Would the significance or non-significance of these values change? __________ Why or why not?

____

26. The following table represents a sample of people who were asked to indicate their occupational category and their city of residence:

City	Professional	Service	Semi-Skilled Labor
		Occupation	
A	20	13	8
B	10	6	12
C	9	23	11

a. First treating city as the independent variable, what is the value of F_{iv}? _____ What is the value of M_{dv}? _____

b. What is the value of lambda? _____ What does lambda mean in this case?

____

c. Using occupation as the independent variable, what is the value of F_{iv}? _____ What is the value of M_{dv}? _____

d. What is the value of lambda? _____ What does lambda mean in this case?

____

e. Which variable better predicts the other? _________________________

27. A researcher asked 35 males and 35 females to rate their attitude toward legalized drug use (for, neutral, or against):

		Attitude	
Gender	For	Neutral	Against
Male	19	12	4
Female	8	11	13
Total	27	23	17

a. Using gender as the independent variable, what is the value of F_{iv}? _____ What is the value of M_{dv}? _____

b. What is the value of lambda? _____ What does lambda mean in this case?

____

c. Does it make sense to use attitude as the independent variable? _____ What is the value of lambda if attitude is used as the independent variable? __________ What does this mean? ____________________________

d. Which variable is of more use in predicting the other? ____________________

Answers

Chapter 1 Why the Social Researcher Uses Statistics

Multiple-Choice

1. a	2. d	3. d	4. a	5. d
6. c	7. b	8. c	9. b	10. b
11. c	12. b	13. a	14. b	15. d
16. a	17. a	18. d	19. b	20. a
21. d	22. c			

Exercises

23. a. amount of machine noise
    b. number of eyes completed

24. a. survey
    b. age
    c. income

25. a. survey

26. a. experiment
    b. mother's depression status
    c. child's reaction to experiment

27. a. experiment
    b. prison rehabilitation programming received
    c. arrest subsequent to release

28. a. experiment
    b. diet
    c. intelligence test score

29. a. content analysis
    b. gender
    c. aggressive language and themes in writing

30. nominal

31. categories are not exhaustive

Chapter 2 Organizing the Data

Multiple-Choice

1. a	2. b	3. b	4. d	5. d
6. c	7. b	8. a	9. d	10. a
11. d	12. b	13. a	14. c	15. c
16. a	17. c	18. b	19. d	20. c
21. a	22. d	23. c	24. b	25. c

Exercises

26.

x	f	cf	%	c%
2	1	1	10	10
3	3	4	30	40
4	1	5	10	50
6	2	7	20	70
9	1	8	10	80
10	2	10	20	100

27. a. 50%
    b. 8

28. a. .48
    b. 10.6%
    c. .16
    d. 8.4%

29. a.

Class Interval	midpoint	f	%
21 –25	23	10	18.52
16 – 20	18	24	44.44
11 – 15	13	13	24.07
6 – 10	8	6	11.11
1 – 5	3	1	1.85
Total		54	100.00

b. 5
c. 16 - 20
d. 3

30. a.

x	f	cf	%	c%
12	1	20	5	100
11	1	19	5	95
10	0	18	0	90
9	3	18	15	90
8	2	15	10	75
7	1	13	5	65
6	2	12	10	60
5	3	10	15	50
4	2	7	10	35
3	2	5	10	25
2	1	3	5	15
1	2	2	10	10
Total	20			

31. a. 35.40%
b. 69.33%
c. .48
d. .65
e. 30.67%
f. .52

32. a. 5
b. 48
c. 35.5 and 30.5
d. 10%
e. 36%

33. a. 14.16%
b. 3.54%
c. 17.40%
d. 39.77%
e. 48.10%
f. 24.74%
g. 24.14%

34. a. 29.81%
b. 29.73%
c. 12.45%
d. 50.94%
e. 7.92%

35. a. 91 – 100
b. 9
c. 12
d. 105.5

36. a. females
b. 23
c. 9 + 3 = 12

d.

Class Interval	f	%	cf	c%
121 - 130	3	4.41	68	100.00
111 - 120	13	19.12	65	95.59
101 - 110	23	33.82	52	76.47
91 - 100	20	29.41	29	42.65
81 - 90	7	10.29	9	13.23
71 - 80	2	2.94	2	2.94
Total	68	100.00		

e. 86.76%

37. a.

Pie Chart of Ethnic Background of Employees (Using Percentages)

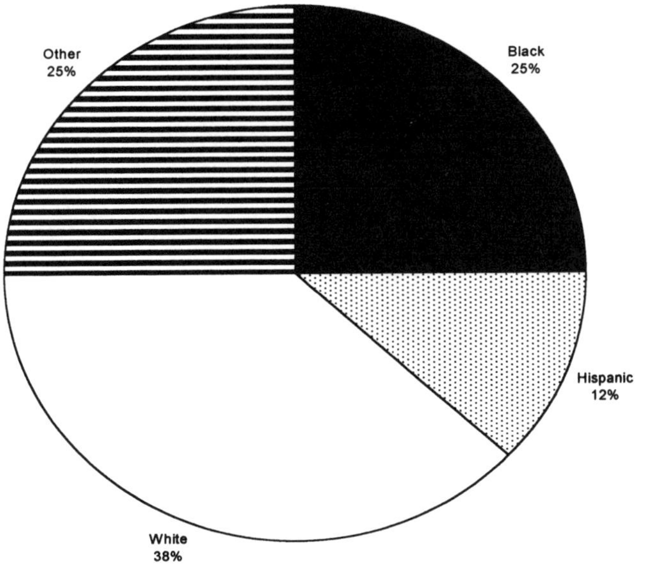

b.

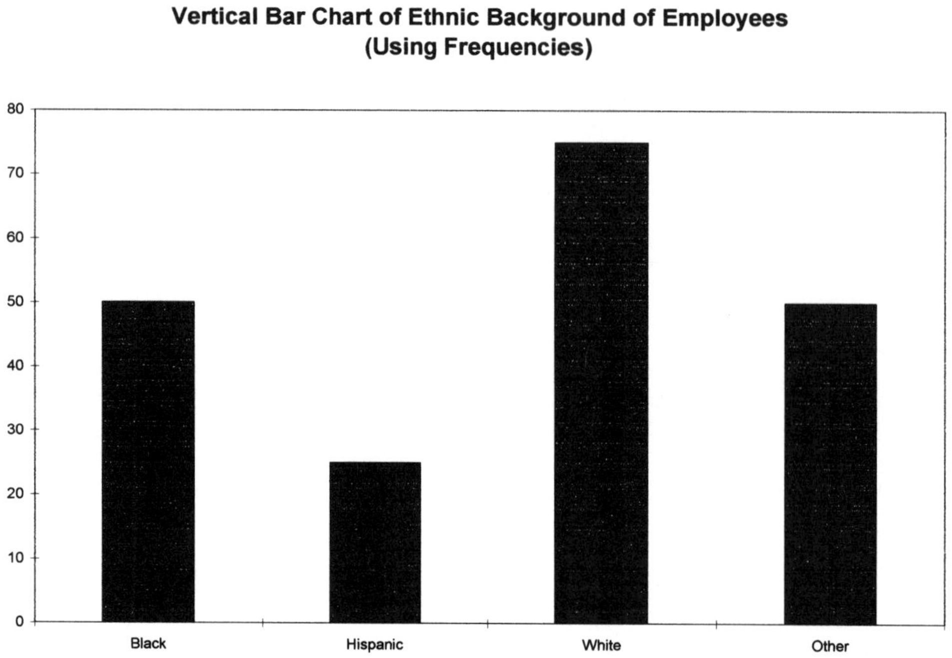

Vertical Bar Chart of Ethnic Background of Employees
(Using Frequencies)
80
70
60
50
40
30
20
10
0
Black
Hispanic
White
Other

38. a.

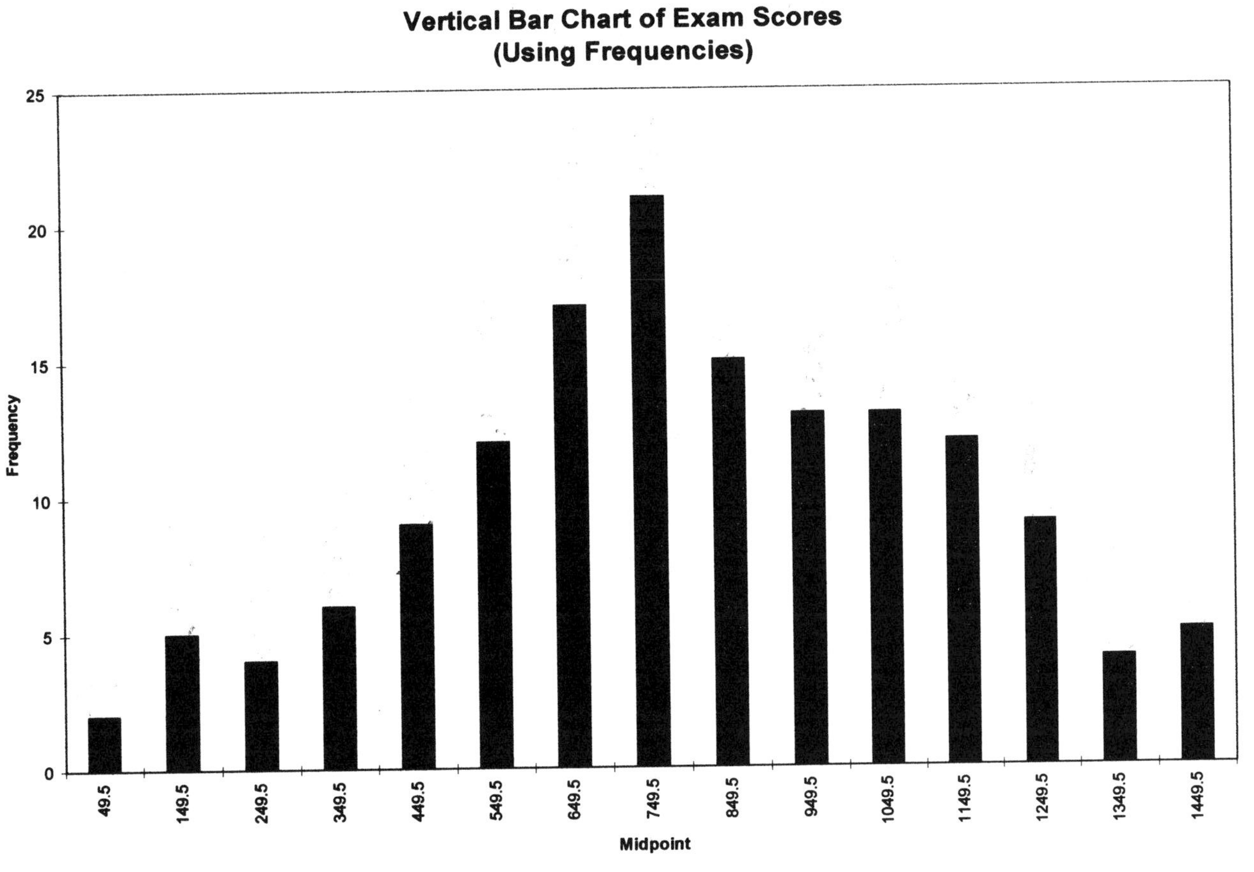

Vertical Bar Chart of Exam Scores
(Using Frequencies)
Frequency
25
20
15
10
5
0
49.5
149.5
249.5
349.5
449.5
549.5
649.5
749.5
849.5
949.5
1049.5
1149.5
1249.5
1349.5
1449.5
Midpoint

b.

Frequency Polygon of Exam Scores

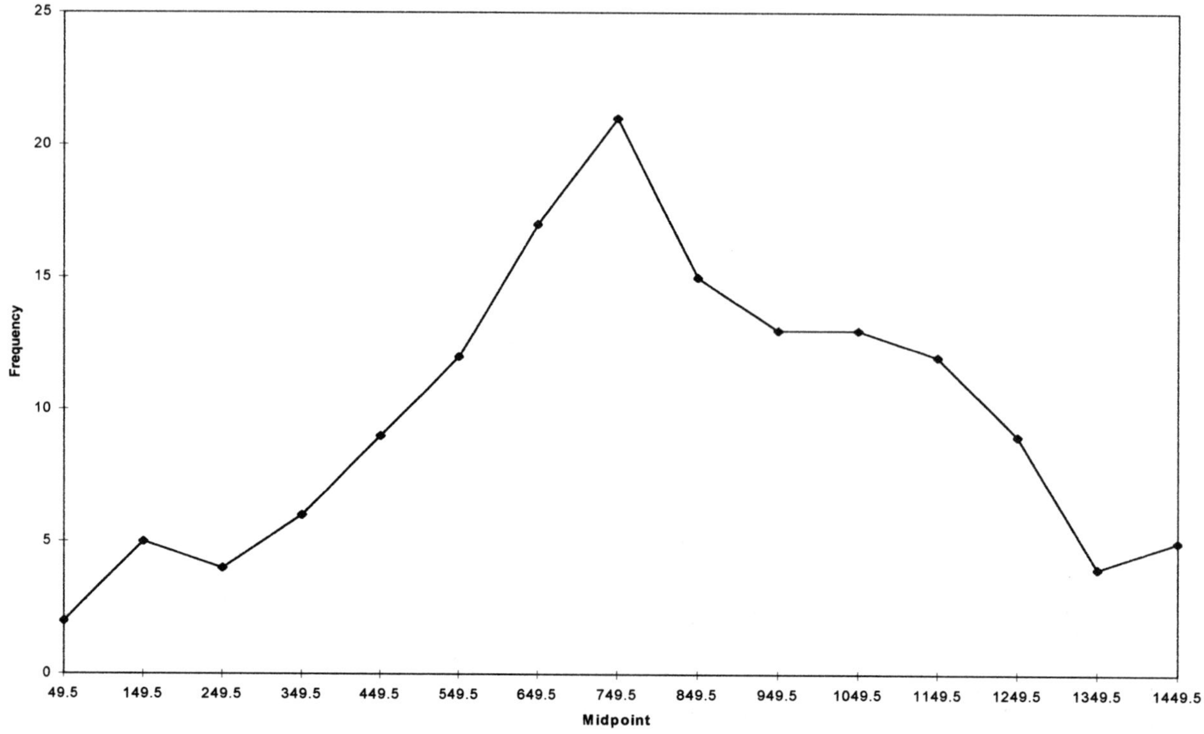

c.

Cumulative Frequency Polygon of Exam Scores

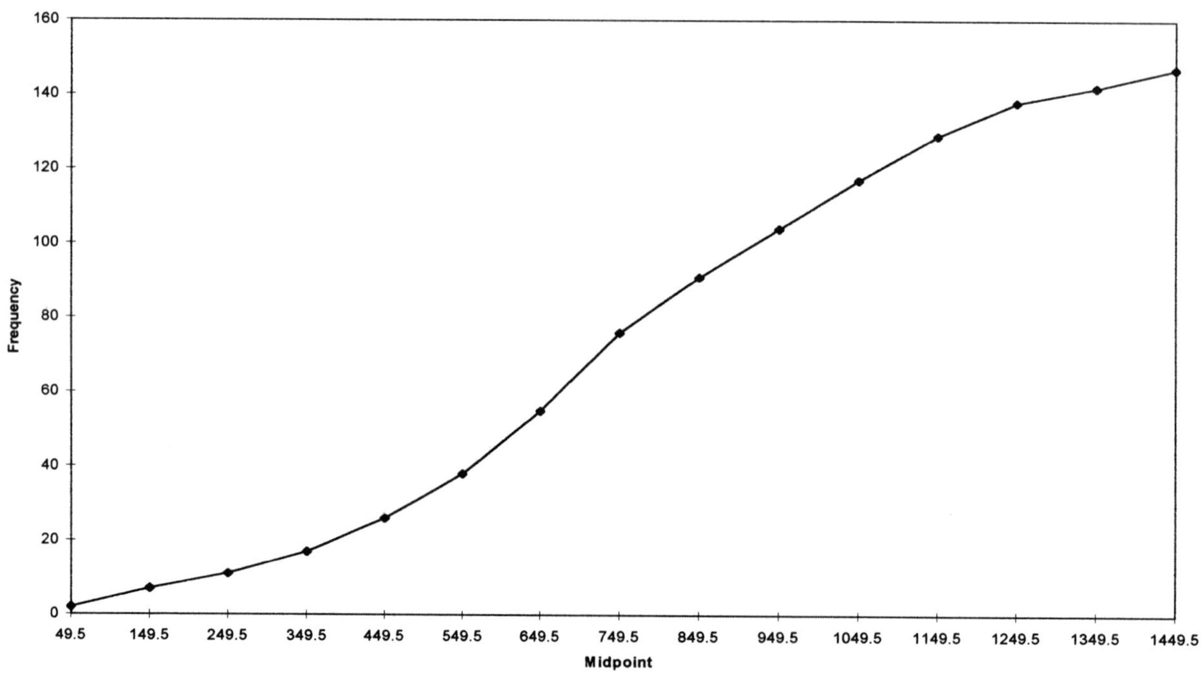

Chapter 3 Measures of Central Tendency

Multiple-Choice

1. d	2. a	3. c	4. a	5. d
6. d	7. c	8. c	9. b	10. b
11. d	12. a	13. d	14. a	15. b
16. d	17. a	18. b	19. b	

Exercises

20. mean, median, mode

21. mode, median, mean

22. all are in the same position

23.
a. 3.4
b. 3.1
c. 3.11

24.
a. 24
b. 18.5
c. 18.25

25.
a. 7
b. 2
c. -1

26.
a. 34.71
b. 29, 33, & 38
c. 35
d. 7.29

27.
a. 10.73
b. -5.27
c. -16.27

28.
a. 55.50
b. 58.28
c. 57.70

29.
a. 31
b. -8
c. 0
d. -17

30.
a. 9 & 10
b. 8
c. 7.22
d. –4.22
e. 1.78

31. a. 1/2 lb.
    b. .46 lbs.

32. a. 2.9
    b. 2.9
    c. 2.98

33. a. 22.46
    b. 22
    c. 1.54

Chapter 4 Measures of Variability

Multiple-Choice

1. d	2. d	3. c	4. b	5. e
6. b	7. c	8. d	9. c	10. b
11. c	12. d	13. c	14. a	15. a
16. c	17. b			

Exercises

18. a. 15
    b. 25.62
    c. 5.06

19. a. 8
    b. 7.41
    c. 2.72

20. a. Males = 6.29; Females = 5.86
    b. Males = 6; Females = 4
    c. Males = 1.47; Females = 1.31
    d. Males = 1.82; Females = 1.45
    e. Males

21. a. 7
    b. 5.94
    c. 2.44

22. 5.68

23. a. 4
    b. 1.39

24. a. 7
    b. 2.29

25. a. 2.83
    b. 2.16

26. a. 6,000
    b. 17,000
    c. Social Policy Research (SPR) = 24,000, Center for Social Research (CSR) = 23,750

d. SPR; since there is less variability, the probability is great that you would make close to $24,000.
e. CSR; while nobody at SPR made over $30,000, 2 of the 8 people studied at CSR made over $30,000; the top salary at CSR is $5,000 higher than the top salary at SPR

27. Jan = 7 Peter = 5

28. Jan = 2.24 Peter = 1.44

29. Jan = 2.48 Peter = 1.85

30. Jan

Chapter 5 Probability and the Normal Curve

Multiple-Choice

1. c	2. c	3. a	4. d	5. c
6. b	7. b	8. c	9. c	10. c
11. b	12. d	13. c	14. b	15. b
16. a	17. b	18. b	19. b	20. c
21. b	22. c	23. d	24. c	25. a
26. c	27. c	28. a	29. c	30. A
31. c	32. a			

33. a. 6/14
b. 11/14
c. 8/14
d. 1 (14/14)

34. .00077, or 1/1,296

35. a. 2/52
b. 26/52
c. 2/52

36. a.

X	Probability (*P*)
2	1/36
3	2/36
4	3/36
5	4/36
6	5/36
7	6/36
8	5/36
9	4/36
10	3/36
11	2/36
12	1/36
Total	______

b. .1667, or 6/36

37. a. -2.31
    b. 3.08
    c. -0.77
    d. 1.15

38. a. 0.63
    b. -0.38
    c. .4481

39. a. 2.33
    b. 3.26
    c. .754
    d. .0099

40. a. .2236
    b. .0038
    c. .2706
    d. .0141

41. a. 18.67
    b. 4.53
    c. .7910
    d. .6141

42. a. 1.74
    b. 0.83
    c. 1.82
    d. 1.52

43. a. .0409
    b. .7967
    c. .9656
    d. .9357

44. Spanish

45. Mathematics

Chapter 6 Samples and Populations

Multiple-Choice

1. c	2. d	3. b	4. b	5. d
6. c	7. a	8. a	9. b	10. b
11. c	12. c	13. b	14. b	15. b
16. a	17. c	18. c	19. f	20. b
21. c	22. b	23. d	24. d	25. c
26. c				

Exercise

27. Your random sample of 15 will differ from this one unless you started at the same point, which was the intersection of column 10 and row 20:

    a. 03 30 45 13 18 41 48 19 39 36 06 26 49 34 07

28. 0.70

29. 41.91

30. 27.58

31.
  a. 1.04
  b. 68% confidence interval = 63.96 to 66.04
  c. 95% confidence interval = 62.96 to 67.04
  d. 99% confidence interval = 62.32 to 67.68

32.
  a. 2.85
  b. 27.11 to 38.29
  c. 84%

33. 437 $\pm$ 38.34

34. 437 $\pm$ 50.46

35.
  a. .87
  b. 60
  c. 2.0
  d. 81.66 to 85.14

36.
  a. 4.04
  b. 0.25
  c. .64
  d. 2.86
  e. 5.51 to 9.19

37. 4.04 $\pm$ 0.49

38. 4.04 $\pm$ 0.65

39.
  a. .03
  b. .6049 to .7032
  c. That there is a 95% likelihood that the true population proportion of people who believe that marijuana should be legalized is neither smaller than .64 nor larger than .70. There is a 5% chance that the true population proportion will be either less than .64 or larger than .70.
  d. No. There is very little chance that, based on our sample data, the percentage of people supporting legalization is any greater than 70%.

Chapter 7 Testing Differences Between Means

Multiple-Choice

1. d	2. d	3. b	4. b	5. b
6. a	7. b	8. d	9. c	10. a
11. d	12. c	13. a	14. b	15. b
16. a	17. e	18. c	19. d	20. a
21. b	22. e	23. d		

Exercises

24. there is no expectation of one mean being greater than the other; the research hypothesis states that one mean will be greater than the other.

25. A distribution resulting from repeatedly taking samples from a single population, and computing the mean of each sample. These sample means are the numbers comprising the distribution.

26. a. 0.74
b. -4.05
c. yes
d. yes

27. a. 2.15
b. 6.98
c. yes
d. yes

28. a. $\bar{X} = 3; \bar{X} = 2$
b. $S_{HS} = 1.22;\ S_{LS} = 1$
c. .6
d. 1.67
e. 14
f. 2.145
g. accept H_o
h. t ratio is used because the true population standard deviations are unknown.

29. 60

30. yes

31. a. $\mu_1 = \mu_2$
b. -36
c. 4.86
d. 2.17
e. critical t = 2.015
f. accept H_o

32. a. $\mu_{BC} \geq \mu_{AC}$
b. 3.96
c. 1.32
d. -2.821
e. accept H_o
f. reject H_o

33. No

34. t = 5.70, critical t = 2.042, statistically significant at $\alpha = .05$

35. a. 1.94
b. 0.73
c. 1.88
d. accept
e. accept

36. yes

37. a. 5.62
b. 0.50
c. no

d. no

38. z = -1.09, critical z = -1.96, not statistically significant

Chapter 8 Analysis of Variance

Multiple-Choice

1. d	2. a	3. b	4. a	5. c
6. d	7. b	8. a	9. c	10. c
11. d	12. a	13. c	14. e	15. a
16. e	17. b	18. d	19. c	20. d

Exercises

21. a. 5.14
b. 2
c. 6.57
d. 109.42
e. 32.91
f. 76.51

22. 4

23.

Source of Variation	df	SS	MS	F
Between Groups	2	62.1	31.05	5.00
Within Groups	6	37.2	6.2	
Total	8	99.3		

24.

Source of Variation	df	SS	MS	F
Between Groups	3	42	14	4
Within Groups	4	14	3.5	
Total	7	56		

25. a. yes; obtained F exceeds critical F
b. no; obtained F does not exceed critical F

26. a. no; obtained F does not exceed critical F
    b. no; obtained F does not exceed critical F

27. a. $\bar{X}_{\text{phone sales}} = 13.13$; $\bar{X}_{\text{coupons}} = 8.63$; $\bar{X}_{\text{direct contact}} = 16.25$
    b. 23.58
    c. 5.78
    d. reject H_o

28. a.

	X_{coupons}	$X_{\text{phone sales}}$	$X_{\text{direct contact}}$
X_{coupons}		4.5	7.62
$X_{\text{phone sales}}$			3.12
$X_{\text{direct contact}}$			

    b. 4.64
    c. 3.66
    d. $X_{\text{coupons vs.}}$ $X_{\text{phone sales}}$, X_{coupons} vs. $X_{\text{direct contact}}$ are significant; $X_{\text{phone sales}}$ vs. $X_{\text{direct contact}}$ are not significant

29. a. accept H_o
    b. No. The F is not significant, so Tukey's HSD is inappropriate.

30. a. F = 1.97, critical F = 3.88, not significant.
    b. Since the F is not significant, Tukey's HSD is inappropriate.

Chapter 9 Nonparametric Tests of Significance

Multiple-Choice

1. a	2. a	3. b	4. a	5. c
6. d	7. b	8. b	9. b	10. d
11. b	12. c	13. d	14. a	15. b
16. b	17. b	18. c	19. d	20. a

Exercises

21. a. 40 in each category.
    b. 15.93
    c. 4
    d. 9.488
    e. reject null hypothesis

22. a. 20
    b. 16.3
    c. 11
    d. 19.675
    e. 24.725
    f. Accept the null hypothesis at both levels

23. a. 21.5
    b. 13.5
    c. 21.5

d. 13.5
e. 2.954
f. 1
g. 6.635
h. Accept H_o

24. a. 5.58
b. 2
c. accept the null hypothesis

25. a. 22.07
b. 6
c. Reject the null hypothesis
d. Reject the null hypothesis

26. a. 6
b. 7
c. 9
d. 0.96
e. accept the null hypothesis

27. a.

Unfavorable Lecture	Rank	Favorable Lecture	Rank
40	15	1	3.5
9	7	12	8
15	9	34	14
0	1.5	25	12
18	11	0	1.5
16	10	46	16
7	6	29	13
3	5		
1	3.5		

b. 23; 40
c. 12
d. There is no difference in self-reported drug use between the two lecture groups
e. Accept H_o

28. a.

G	Rank	PG	Rank	PG-13	Rank	R	Rank	NC-17	Rank	XXX	Rank
1	9.5	5	25	2	15	10	33	12	35	0	3.5
0	3.5	2	15	8	30	15	36	6	26.5	2	15
2	15	3	19	4	22.5	4	22.5	1	9.5	0	3.5
0	3.5	1	9.5	3	19	8	30	4	22.5	1	9.5
1	9.5	4	22.5	7	28	11	34	8	30	1	9.5
0	3.5	3	19	6	26.5	9	32	2	15	0	3.5

b. 187.5
c. 24.42
d. 5
e. 11.07
f. Reject H_o

29. a. 160
b. 9; 9
c. 3.281
d. 3.841
e. No significant gender differences exist in the amount charged to males versus females for repair work.
f. Reject H_o

Chapter 10 Correlation

Multiple-Choice

1. b	2. c	3. a	4. b	5. c
6. b	7. d	8. c	9. a	10. b

11. d	12. a	13. b	14. d	15. c
16. a	17. c	18. d	19. d	20. a
21. c	22. d			

23. a. 17.2; 68.8
    b. 34.4
    c. 1.0

24. a 53.52; 55.52
    b. -50.5
    c. -.9264

25. a. 6.8
    b. no
    c. no

26. a. age & debt and age & salary
    b. age & salary
    c. .5010
    d. greater than

27. a. .4494
    b. .6319
    c. no
    d. .7646
    e. no

28. a. .8526
    b. .6319
    c. .7646
    d. yes; both levels
    e. positive

29. a. .26
    b. negative
    c. .98
    d. .12
    e. .30

30. a. -.69
    b. .7977
    c. no
    d. negative

31. a. .90
    b. yes
    c. Without controlling for number of friends, the relationship between IQ and years of employment is suppressed (-.69 is significant at the .05 level, but not at the .01 level).

32. a. .9614
    b. .6319
    c. yes

33. a. .55
    b. .77
    c. The correlation between smoking and heart disease is stronger when controlling for age.

Chapter 11 Regression Analysis

Multiple-Choice

1. b	2. c	3. c	4. a	5. b
6. b	7. b	8. c	9. a	10. b
11. d	12. b	13. b	14. d	15. a
16. d	17. a	18. c	19. a	

Exercises

20.
- a. 50.61
- b. 38.46

21.
- a. -.62
- b. 5.63
- c. 2.53
- d. 3.77

22.
- a. 1618; 99; 277
- b. 252.66; 18.92; -53.66
- c. -.21
- d. 4.85
- e. Y = 4.85 - (.21)X
- f. 4.85, or about 5
- g. .21 (so, for every 5 years of education, you would expect one less child)

23.
- a. 9.40
- b. 3.24

24.
- a. 1.375
- b. 17.1786
- c. 25.4286

25.
- a. no, because there is no way of knowing whether it is positive or negative
- b. .4%
- c. 4.18
- d. .10; no

26.
- a. 85%
- b. 62%; 33%

27. a.

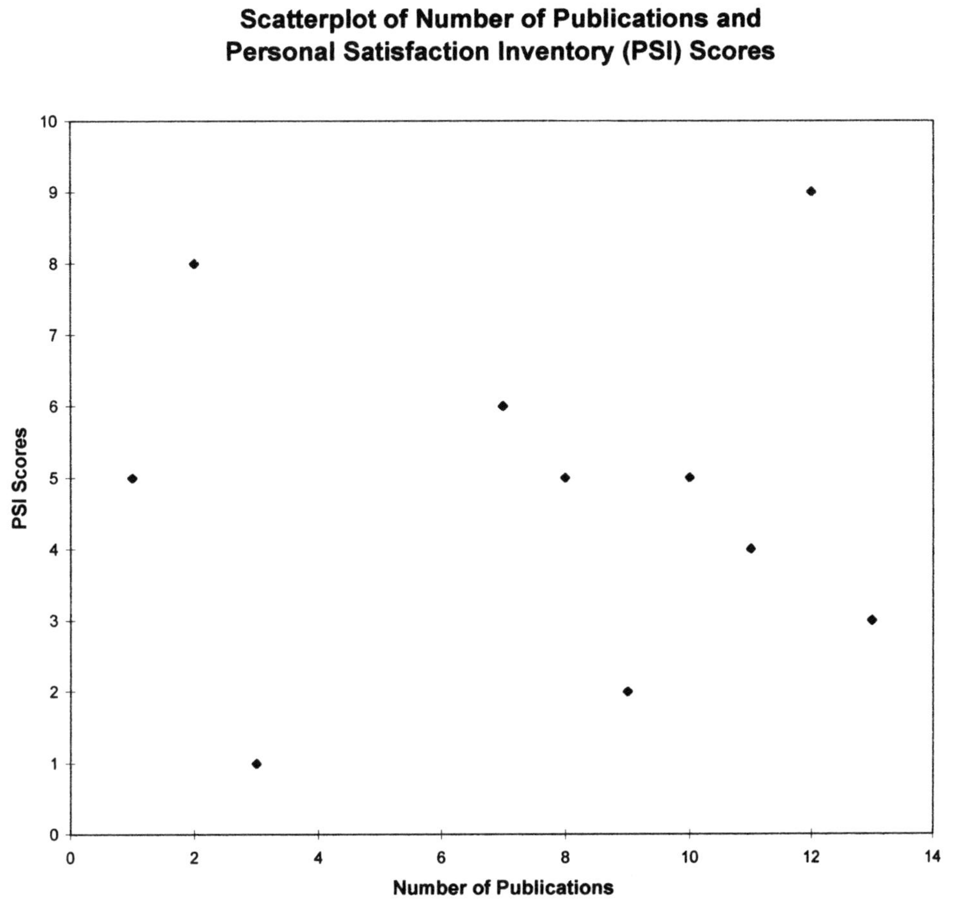

b. There doesn't appear to be much of a correlation.
c. 0 (.0012)
d. 4.79

e.

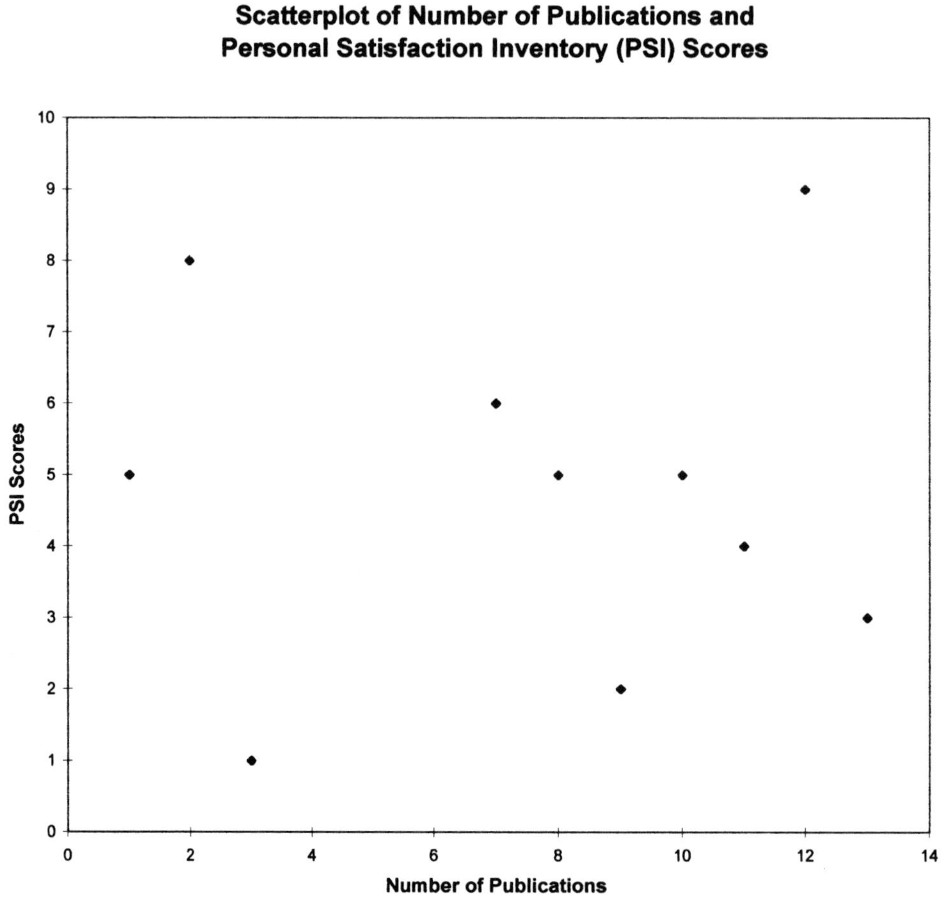

f. 0 (.0021)

g. 4.79; No - since there is virtually no correlation between PSI and publications, the values of PSI are essentially the same across the values of the publication variable.

Chapter 12 Nonparametric Measures of Correlation

Multiple-Choice

1. c	2. d	3. b	4. c	5. c
6. a	7. b	8. a	9. b	10. b
11. c	12. a	13. b	14. d	15. b

Exercises

16. 0.345

17.
a. 54
b. 0.673
c. reject the null hypothesis at the .05 level of significance
accept the null hypothesis at the .01 level of significance

18.
a. .988
b. .475
c. .625
d. reject the null hypothesis at both levels

19.
a. .667
b. .648
c. .794
d. reject the null hypothesis at the .05 level of significance; accept at the .01 level

20.
a. 1800
b. 2488

21.
a. -0.1604 (weak negative correlation)
b. reject the null hypothesis at the .05 level of significance
c. accept the null hypothesis at the .01 level of significance

22.
a. 2.35
b. .19
c. 1
d. 3.841
e. There is no association between prison drug treatment and arrest subsequent to release from prison (fail to reject H_o).

23.
a. 1.28
b. .16
c. There is no statistically significant relationship between gender and smoking (fail to reject H_o).

24.
a. 29.27
b. .45 (contingency coefficient, because the table is larger than 2 x 2)
c. 12.592
d. yes
e. no; a square table is better for calculating the contingency coefficient (the maximum value of the contingency coefficient changes as the rows and columns of a table change, especially for non-square tables).
f. .36
g. yes
h. both C and V show a moderate degree of association between the two variables, but V probably better reflects the data, since this is a non-square table.

25. a. .36; yes
    b. .28; yes
    c. C would increase; V would increase; no; the significance is tested by testing the obtained chi-square, and this significance is dependent on the degrees of freedom (and α).

26. a. 55; 43
    b. .17; using city as the independent variable, we reduce the error of our prediction by .17
    c. 55; 42
    d. .19; using occupation as the independent variable, we reduce the error of our prediction by .19
    e. Occupation more accurately predicts city of residence than vice versa.

27. a. 32; 27
    b. .13; using gender as the independent variable, we reduce the error of our prediction by .13
    c. yes; .28; if you know a person's attitude toward drug use, you can predict their gender about 28% correctly (i.e., if lambda were 1.0 in this case, knowing a person's attitude would accurately predict that person's gender 100% of the time)
    d. attitude predicts gender better than gender predicts attitude

NOTES

NOTES

NOTES

NOTES

NOTES

NOTES

NOTES

NOTES